... extinction updates essential genetic modules. The situational change of asteroid impact led to the necessity for smaller body size. Most dinosaurs could not fulfill that necessity, mammals and crocodiles could. Both clades were of comparable adaptation levels. Regarding adaptability, however, mammals turned out to be more 'creative' ...

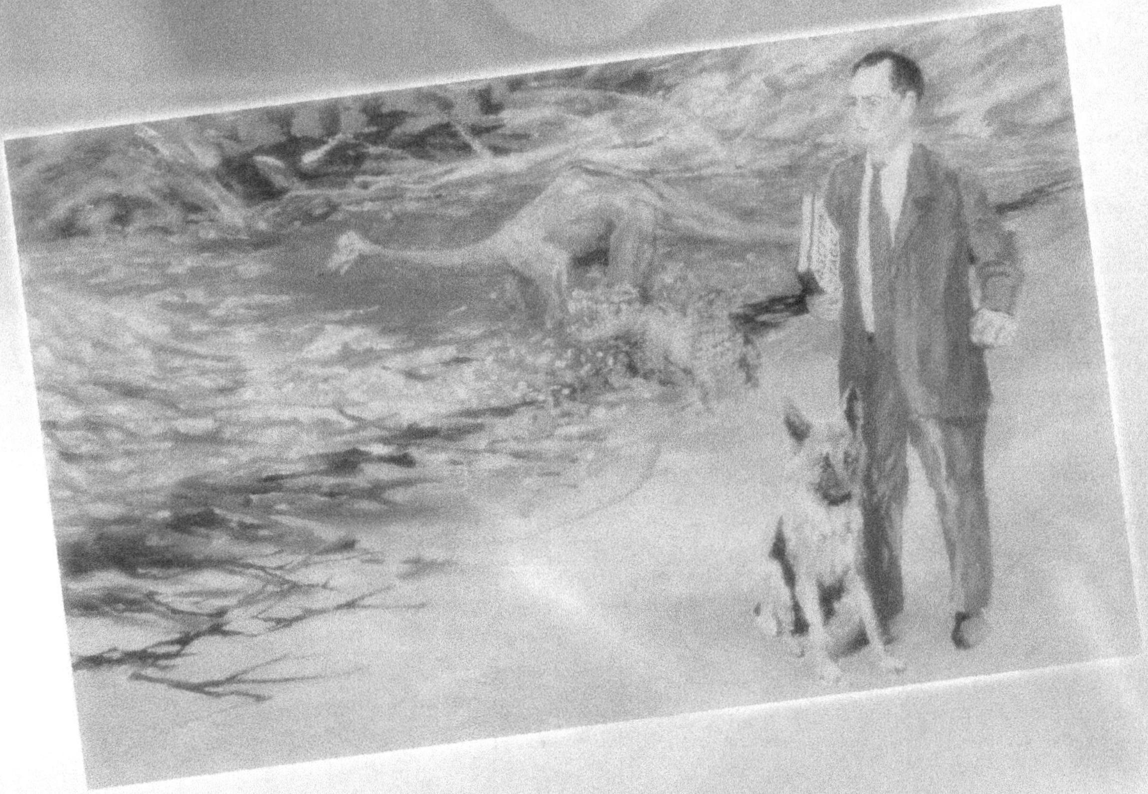

Dr. Thomas F. Kochmann has been working for many years in the field of mathematical modelling for his company Creative-Agents.net e.K in Germany.

For many decades, a controversial discussion has smouldered about the question: What is the unit of Darwinian selection? Versatile answers have aroused among evolutionary biologists. Some attributed this role to the „egoistic gene" unit only (e.g. Richard Dawkins), others paid additional attention to individual organisms or groups of individual organisms (e.g. David Sloan Wilson, Elliot Sober), or even species or clade selection (e.g. Stephen Jay Gould, Niles Eldredge), ultimately leading to hierarchical principles such as multilevel-selection. Attempts to address this question mathematically have been performed, however mainly using tools and techniques of the general mathematical field „Analysis" (e.g. by stochastic or deterministic ordinary or partial differential equations). Contrary to all of those the author started a different approach, namely incorporating the mathematical field of „Logic" into the scenary, constructing a formal evolutionary framework. Many new expressions such as „necessity" and „adaptive reaction" are introduced and require some practise from the reader. For this reason a multiplicity of (molecular) biological examples is given in order to deepen their usage and understanding. In the end, they will enable the reader to turn around the original bottom-up „imperative" view into a much more abstract top-down „declarative" perspective, where „evolution" could be regarded as a generalized satisfiability problem of an arbitrary logical formula. This reprint of the original work from 2001 is dedicated to Stephen Jay Gould (1941-2002).

Bibliographic information published by the German National Library:
The German National Library lists this publication in the Deutsche Nationalbibliografie: detailed bibliographic data are available on the Internet at http://dnb.dnb.de.

© 2001 Thomas F. Kochmann, unchanged reprint 2019

Production and publisher: BoD – Books on Demand, Norderstedt
ISBN: 978-3-7481-7360-1

LOGICAL EVOLUTIONARY LANGUAGE (LEL)

A Mathematical Model for Darwinian Selection on Multiple Levels

Thomas F. Kochmann

Creative-Agents.net e.K.
49134 Wallenhorst
Germany
thomas.kochmann@creative-agents.net
tkochmann@acm.org

KEY WORDS
- genetic drift - genetic routeing - situational change - necessity - adaptive reaction -
RUNNING HEAD
Mathematical Multi-Level Evolution Model

this work from the year 2001 is dedicated to Stephen Jay Gould

Mathematical Multi-Level Evolution Model

ABSTRACT

The current debate on evolutionary theory revolves around the concept of selection on many levels. We propose an uncertainty principle to be the causative mechanism when evolution has to decide between mutually exclusive paths. We simulated evolutionary incompatibility, *i.e.* a case where more than two phenotypical attributes could not evolve simultaneously in one organism. In some experiments, a compromise between two or more phenotypical attributes evolved which we term heterosynthesis. A biological example of a heterosynthesis is the dual function of aconitase in the citric acid cycle and iron metabolism in triblastians. In other experiments, homosynthesis, the opposite of heterosynthesis, evolved, *i.e.* only one phenotypical attribute without any compromise. We proved mathematically that this homosynthesis resulted from the random selection of one phenotypical attribute. In contrast to classical genetic drift, this randomness takes effect before a phenotypical attribute has been evolved. It signifies the randomness at a branching point during the evolution of species. Therefore, it could be regarded as the randomness on a higher level. Further mathematical analysis revealed an uncertainty principle in Darwin's natural selection. This means that the fittest phenotype does not necessarily win the struggle for existence, *i.e.* it only has a higher probability of surviving.

INTRODUCTION

It is generally assumed that all living species on Earth derived from their progenitors like branches of a multidimensional tree from its trunk (DAWKINS 1988), each branch symbolizing a clade, each leaf symbolizing a species. A species is adapted to its ecological niche (HUTCHINSON 1957) through a characteristic set of phenotypical attributes. As a phenotypical attribute we regard any kind of biological function such as an anatomical structure or a biochemical pathway. How does a species develop a new phenotypical attribute and evolve from one niche to another?

GOULD and ELDREDGE (e.g. 1993) proposed that besides Darwin's organismic selection, another evolutionary mechanism was at work: species selection. Furthermore, GOULD and LLOYD (1999) assumed many levels of Darwinian selection: e.g tissue, organism, deme, species, genus. In order to investigate whether these different selection levels exist and how they interact, we have developed a *logical evolutionary language* which integrates Darwinian selection on many levels as a mathematical model. The expressions of this formal language are precisely formulated in the APPENDIX I. In the main text of this paper, however, these expressions are only informally defined and clarified by molecular biological examples. Here, our formal language is principally used to formulate questions about the randomness in the diversification of species; these questions are answered through computational experiments and further stochastic analysis.

Generally, our formal evolutionary language may serve as a basis for functional simulation of molecular biological and biochemical networks in future projects. With the help of this tool a logical ordering principle of molecular biological and biochemical data banks could be provided. This algorithm may enable such data banks to be intelligent, *i.e.* to draw integrative conclusions from their inputs.

FORMAL EVOLUTIONARY LANGUAGE

Transformation within an ecological niche: The evolution of a species in an ecological niche displays two alternative modes: unidirectional transformation or multidirectional diversification. A reason for transformation is a *situational change* such as a shift in climate. Hence, a situational change may be a changed *external condition* in an ecological niche. Moreover, a situational change can also be an evolved *phenotypical attribute* in organisms inhabiting an ecological niche. An example is a changed colour of petals or the expression of a new protein in iron metabolism.

A situational change can trigger one or more *necessities* for further phenotypical attributes. Necessity is a Lamarckian expression (LAMARCK 1809). However, the flaw in Lamarck's evolutionary theory was that he believed an acquired trait to be inheritable. Nevertheless, we choose the term necessity to verify whether an undeveloped phenotypical attribute would lead to a higher *fitness*. We define the fitness of a population as the expectation (BRÉMAUD, P., 1999) of the net reproduction rates of its organisms. Such a stochastic approach is required, since the reproduction rates may fluctuate and the organisms may be genotypically different.

The fitness f_n of a population containing n organisms is its expected increase in organism count per fixed generation time Δt: $f_n = \frac{b_n - d_n}{\Delta t}$, b_n being its expected birth count, d_n being its expected death count during Δt. The fitness $f = f_1$ of an organism is the fitness of a population containing only this organisms, *i.e.* the expected number of surviving progeny per fixed generation time Δt: $f = \frac{b - d}{\Delta t}$, $b = b_1$ being the expected number of births, $d = d_1$ being the expected number of deaths during Δt. Let $f(a)$ or $f(a')$ be the fitness of an organism which carries or does not carry the phenotypical attribute a. When organisms without a live in a population, a necessity for a prevails if and only if $f(a) > f(a')$.

Mathematical Multi-Level Evolution Model

The necessity for a new phenotypical attribute could trigger one or more *adaptive reactions*. We define an adaptive reaction as the evolution of a new *genotypical instruction*. As a genotypical instruction we regard any change within a genotype, *e.g.* a mutation or a sequence of mutations. Here, a single mutation can be viewed as a *random adaptive reaction*. A sequence of mutations through Darwin's natural selection mechanism (DARWIN 1859) can be viewed as a *nonrandom adaptive reaction*. Generally, we term an adaptive reaction nonrandom when it is caused by a necessity, and we term an adaptive reaction random when it is not caused by any necessity.

Two adaptive reactions ρ_1 and ρ_2 are logically connected by *AND* when they can concurrently occur in one organism, *i.e.* $\rho_1 \wedge \rho_2$. They are logically connected by *XOR* when either ρ_1 or ρ_2 can occur in one organism, *i.e.* $\rho_1 \oplus \rho_2$. They are logically connected by *OR* when either ρ_1 or ρ_1 or both can occur in one organism, *i.e.* $\rho_1 \vee \rho_1$. In the cases of OR and XOR, a random selection of the corresponding adaptive reactions has to take place (before they start to evolve); we term this randomness *genetic routeing*. Thus, genetic routeing 'decides' which adaptive reactions satisfy a given necessity, and which adaptive reactions do not.

A necessity can be illustrated as the peaks of a mountain range attracting a group of climbers who may select one or more peaks to be conquered simultaneously. The peaks of a mountain range in Wright's phenotypical adaptive landscape (WRIGHT 1932, 1982a and 1988) can be visualized as a necessity to a population of organisms. The ascent to one of the adaptive peaks would be an adaptive reaction. Would the population select one or more peaks to ascend simultaneously?

Mathematical Multi-Level Evolution Model

In a geographical landscape, some mountains may be too difficult to be conquered by a group of climbers. Similarly, in an adaptive landscape some adaptive reactions may not be developed by a population of organisms with limited adaptability (RADMAN 1999). These adaptive reactions are infeasible. Necessities unable to be satisfied by feasible adaptive reactions may cause the extinction of a population. Simultaneous extinctions of many populations might cause the extinction of a clade. For instance, the situational change of asteroid impact, the opening of the Atlantic ocean, increased vulcanism, and seasonal aridity 65 million years ago may have triggered a necessity for smaller body size (SERENO 1999). Most dinosaurs – except the ancestors of present-day birds – were unable to develop adaptive reactions for this necessity; they died out. Other vertebrates such as the progenitors of present-day mammals, crocodiles, and tortoises had already developed such adaptive reactions; they survived.

After dinosaurs had vacated their ecological niches, terrestrial ecological niches similar to the vacated ones were successively occupied by mammals rather than reptiles. One reason for the overall success of mammals may have been their greater capability to diversify as a result of their greater adaptability (SERENO 1999). The greater adaptability of a clade is a higher overall probability of adaptive reactions being evolved, *i.e.* a higher overall probability of necessities being fulfilled. As a measure of this adaptability, Gould proposed 'emergent fitness' (*e.g.* GOULD and LLOYD 1999); we term it *species-level fitness*. Thus, we define the species-level fitness ϕ_n of a clade containing n species as its expected increase in species count per fixed life span $\Delta\tau$: $\phi_n = \frac{\beta_n - \delta_n}{\Delta\tau}$, β_n being its expected origination count, δ_n being its expected extinction count. The species-level fitness $\phi_1 = \phi$ of a single species is the species-level fitness of a clade containing only this species, *i.e.* its expected number of surviving progeny species per fixed life span $\Delta\tau$. $\phi=\dfrac{\beta-\delta}{\Delta\tau}$, $\beta=\beta_1$ being the expected number of originations and $\delta=\delta_1$ the expected number of extinctions during $\Delta\tau$.

On the level of a species, *species-level situational changes*, *species-level necessities*, and *species-level adaptive reactions* occur and interact with one another. Here, a *species-level population* is a clade of species, and a *species-level organism* is a single *species*. Likewise, other selection levels can be described, *e.g.* a cancer cell within a tissue or a transposon within a cell genome. For instance, on the level of a transposon, there might be *transposon-level situational changes*, *transposon-level necessities*, *transposon-level adaptive reactions*, and a *transposon-level fitness*. (Transposons are intragenomic parasites which primarily fufil their own evolutionary task. We call this task: 'to replicate as much as possible within one genome' Contrary to viruses, transposons never leave their host.)

An example of a selection level which is lower than the transposon level is an individual nucleotide at a genomic DNA position. Here, the replicase at this DNA position could lead to a *nucleotide-level necessity* for the incorporation of either A, C, T or G. Each possible nucleotide incorporation can be regarded as a *nucleotide-level adaptive reaction*, each of which having a specific probability p, *e.g.* $p(A)=0.99$, $p(C)=0.002$, $p(G)=0.005$, $p(T)=0.003$. Thus, a random selection of either A, C, G or T occurs, *i.e. organism-level genetic routeing*. A point mutation at this nucleotide position can be regarded as 'combined organism level genetic routeing'. We call it: 'either C or G or T'. On the nucleotide level, this adaptive reaction is caused by the necessity for nucleotide incorporation. Thus, on the nucleotide level, this adaptive reaction is nonrandom. On the organism level, however, a point mutation is a random adaptive reaction, *i.e.* it is not caused by an organism-level necessity.

Consequently, it depends on the level of selection whether an adaptive reaction is nonrandom or random. An example of such different perspectives from the species and the organism level is the adaptive reaction of a new biochemical pathway: On the organism level, such an adaptive reaction is (usually) nonrandom, since it is caused by an organism-level necessity. On the species level, however, such an adaptive reaction is random, since it is (usually) not caused by a species-level necessity. From the perspective of the species level, the random selection of an adaptive reaction is genetic routeing, *i.e.* genetic routeing can be regarded as a *random species-level adaptive reaction*, *i.e* as a *species-level mutation*.

On each selection level, overlapping functional units (or semantic units) exist. They result from nonrandom adaptive reactions which are caused by a necessity. This necessity is triggered by a situational change. We term such a functional unit *module* (see precise mathematical Definition 12 in APPENDIX I). A combination of modules is also a module, *i.e.* a *synthesis* of modules. We distinguish two kinds of syntheses: *heterosynthesis* and *homosynthesis* (see Definition 15 in APPENDIX I). A heterosynthesis is a compromise between *incompatible* adaptive reactions, *i.e.* it consists of necessities whose corresponding adaptive reactions are mutually exclusive. An example of such a mutual exclusivity are the digging or catching abilities of the forelimbs in some insects. The adaptive reaction of digging and catching could not develop simultaneously in particular insects; *i.e.* some insects only evolved the digging ability, others only the catching ability. In cases of compromise, *i.e.* heterosynthesis, those necessities could only be partially satisfied. A heterosynthesis implies that the corresponding adaptive reaction consist of *adaptive subreactions* which are categorically different, *i.e.* 'hetero'.

Mathematical Multi-Level Evolution Model

In a homosynthesis, the corresponding adaptive reactions are categorically similar, *i.e.* 'homo'. Thus, a homosynthesis only consists of necessities whose adaptive reactions are not mutually exclusive. Hence, these necessities can be completely satisfied, *i.e.* they do not require a compromise. An example of a homosynthesis is a protein which consists of several interacting (or cooperating) domains. These domains are not mutually exclusive since they occur and function simultaneously. Furthermore, a single functional domain of a protein could also be regarded as a homosynthesis, since its corresponding adaptive reaction cannot be divided in a sequence of mutually exclusive adaptive subreactions.

An example of heterosynthesis is the protein cytoplasmic aconitase in higher eukaryotes, conceivably descended from an ancient cytoplasmic citric acid cycle enzyme (KREBS 1970; HUYNEN *et al.* 1999). This module consists of two submodules: 'citrate transformer' and 'translation regulator', whose biological functions are mutually exclusive (HENTZE and KUEHN 1996; GEHRING *et al.* 1999). The citrate transformer enzymatically converts citrate into isocitrate (KREBS 1970); the translation regulator (also called iron regulatory protein-1 or IRP-1) impedes gene expression of the iron-storage protein ferritin by binding to the 5' untranslated region (5'UTR) of its m-RNA (Hentze *et al.* 1989). If the ratio of cytoplasmic iron to oxygen concentration decreases, cytoplasmic aconitase loses an iron-sulphur cluster switching the citrate transformer off and the translation regulator on. This 'toggle switch' can be reversed by increasing the ratio of cytoplasmic iron to oxygen concentration (GEHRING *et al.* 1999).

Both submodules have been found coexistent in vertebrates, *Drosophila melanogaster*, and *Caenorhabditis elegans* (ROUAULT and KLAUSNER 1997; MUCKENTHALER *et al.* 1998) – all these clades are triblastians. 'Lower' clades such as bacteria and yeast do not necessarily carry the cytoplasmic aconitase gene (HUYNEN *et al*. 1999). However, in cases where they do only the citrate transformer has been detected so far, not the translation regulator (ROUAULT and KLAUSNER 1997; MUCKENTHALER *et al*. 1998).

Thus, we propose that the citrate transformer is the phylogenetically 'older' submodule. We suggest that its situational change was the increased atmospheric oxygen concentration prior to the Cambrian explosion (OHNO 1997). The necessity of the citrate transformer was an intensified generation of energy through oxidative metabolism. Its adaptive reaction was the evolution of the enzymatic citrate specificity.

The translation regulator could arguably be the phylogenetically 'younger' submodule. Then, its situational change was multicellularity in early triblastians, *i.e.* a larger mass-surface ratio and a higher oxidative metabolism rate. The necessity of the translation regulator was an intensified oxygen sequestration into specialized transporting and detoxifying iron proteins (HENTZE and KUEHN 1996; GEHRING *et al*. 1999). Its adaptive reaction was the evolution of the 5'UTR binding specificity.

The unusual feature of this evolution is that while a new adaptive reaction was developing, the result of a previously completed adaptive reaction was preserved in the same protein (GEHRING *et al*. 1999; PARASKEVA, and HENTZE 1996; Narahari *et al*. 2000). Therefore, it is likely that the necessity for the established citrate transformer activity was also preserved. Why was the necessity for the translation regulator not satisfied through the development of an additional module, *e.g.* a new protein? And why was the preadapted module to satisfy this necessity cytoplasmic aconitase? Was its selection random or nonrandom (*i.e.* was its selection due to a random or a nonrandom adaptive reaction)?

Questions like these also apply to homosyntheses such as the genetic code. It seems plausible that its situational change was the availability of amino acids and RNA molecules after the completion of chemical evolution (ORGEL 1998). Its necessity was one combination of 64 codons with 20 amino acids, and translational stop and start trigger mechanisms. From basic combinatorical reasoning we can show that there would have been $\frac{64!}{42!} \times 22^{42} \approx 2.175 \times 10^{94}$ possible adaptive reactions to satisfy that necessity. Why did only one become universal? For example, why is AUG the start codon and not any other? Moreover, why does it code for methionine in eucaryotes or N-formyl-methionine in procaryotes? Was nature's selection of AUG random or nonrandom (*i.e.* due to random or nonrandom adaptive reactions)? No evidence for nonrandomness (*i.e.* nonrandom adaptive reactions) has yet been found (KNIGHT *et al.* 1999).

Therefore, we hypothesize that the necessity to develop a start codon would have had 64 adaptive reactions leading to an (approximately) equal fitness. If the simultaneous occurrence of more than one adaptive reaction within an organism neither reduced nor increased its overall fitness, the adaptive reactions would be *isoadditive*. If the simultaneous occurrence reduced an organism's overall fitness, the adaptive reactions would be *hypoadditive*. It seems plausible that two hypoadditive adaptive reactions are logically connected by XOR. Thus, if the 64 possible adaptive reactions which were able to satisfy the necessity 'start codon' had been hypoadditive, only one of them would have been able to evolve, *i.e.* either AAA or AAC or AAG or ... or AUG or ... or UUG or UUU.

If the simultaneous occurrence of adaptive reactions increased an organism's overall fitnesss, they would be *hyperadditive*. This occurs when each individual adaptive reaction satisfies a categorically different necessity. For example, in an ecological niche of a bacterial population four antibiotics are administered simultaneously. The dosages are sufficiently low to prevent the extinction of this population. This situational change triggers a set of four categorically different necessities, *i.e.* resistance to each antibiotic. The corresponding four adaptive reactions are hyperadditive. If they are logically connected by AND, *i.e.* if the corresponding necessities can be satisfied concurrently, the species transforms unidirectionally. We term this ability *compatibility*, *i.e.* all adaptive reactions are *compatible* with each other. However, what happens when only three necessities or less can be satisfied concurrently?

Diversification within an ecological niche. A species may diversify when several necessities cannot be satisfied simultaneously. We call this inability *incompatibility*, *i.e.* the corresponding adaptive reactions are incompatible with each other. One example is disruptive selection (MAYNARD SMITH 1966; FELSENSTEIN 1981; KONDRASHOV 1986; DIEHL and BUSH 1989), *i.e.* when the corresponding adaptive reactions are hypoadditive. Another example is genomic restriction, *i.e.* when the corresponding adaptive reactions are hyperadditive and genetic 'space', required for their simultaneous evolution, is lacking. We distinguish two types of genomic restriction, *i.e. genomic incompatibility*: *structural incompatibility* and *functional incompatibility*.

Structural incompatibility means that the complete genome is too small to satisfy all necessities simultaneously. Polyhedral viruses such as the lambda phage (LEWIN 1997) are good examples. Their genomes are not able to extend, as they have to be packed into pseudospherical protein casings. Contrary to this, other clades, *i.e.* helical viruses, procaryotes, eucaryotes, were able to extend their genomes. This ability to extend can be regarded as a species-level adaptive reaction fulfilling the species-level necessity of preventing structural incompatibility; such a species-level adaptive reaction may be one reason for genetic redundancy. (An example of such a redundancy are eucaryotic genomes consisting of a major part of heterochromatin exhibiting no known biochemical function.)

In clades with redundant genome structures (such as triblastians) duplications of preadapted modules may occur. In an undifferentiated somatic tumor cell duplicated copies may be hyperadditive and logically connected by AND causing it to grow aggressively (BRODEUR and HOGARTY 1998). In a germ line cell, however, an isoadditive relationship between duplicated copies would mutationally destabilize most of them, *e.g.* leading to the creation of pseudogenes (MIGHELL *et al.* 2000). Nevertheless, a new situational change can make some copies in a germ line become precursors for further hyperadditive and AND connected adaptive reactions (BROSIUS and GOULD 1992). (This means that such a situational change can trigger the individual evolution of those copies.) There is functional incompatibility when too few module copies are available to satisfy a given number of necessities simultaneously.

Incompatibility within a species may lead to subpopulations fulfilling diverse subsets of necessities in an ecological niche. This results in the occupation of diverse subniches. The organisms of such subniches differ slightly in strategies of resource utilization or coping with harmful conditions. If they induce subsequent situational changes, further necessities may arise, possibly triggering further transformation or diversification within the subniches, and intensifying isolation of the subpopulations.

When a threshold degree of isolation has been reached, *e.g.* mating incongruity in sexual species, diverse species will evolve by sympatric speciation (MAYR 1942 and 1992;

MAYNARD SMITH 1966; FELSENSTEIN 1981; KONDRASHOV 1986; DIEHL and BUSH 1989). In sympatric speciation isolation develops simultaneously with the speciation process, whereas in allopatric speciation isolation is the primordial situational change (MAYR 1963).

METHODS AND RESULTS

Allopatric speciation simulation: Let a homogeneous population of a species live in a habitat of an ecological niche. When geographic isolation divides the habitat into different subhabitats with similar subpopulations a set of analogous necessities prevails in each subhabitat. Allopatric speciation occurs where different adaptive reactions develop in different subhabitats. However, this has almost exclusively been described in small populations of sexual species (MAYR 1963). Using an appropriate algorithm implemented in the programming language C++ we examined its occurrence in large populations of an asexual species.

In our computer experiments, a population comprised 1,000 organisms which occupied a subhabitat of a virtual ecological niche. Their DNA sequences at the start were generated randomly, each consisting of 100 nucleotides. (Annotation 1) In the ecological niche, one necessity prevailed, which was able to be satisfied by four different isoadditive adaptive reactions leading to equal fitness.

In order to specify the four adaptive reactions we term the peaks of their representative adaptive mountains *target sequences*. The number of nucleotide positions where two sequences differ is known as Hamming distance (HAMMING 1980). We define the average of all Hamming distances to all target sequence combination pairs in a target sequence set as its *diversity degree*.

100 experiments were carried out where the target sequence set had diversity degrees 1 to 100. During each generation the organisms replicated according to their individual fitness, some died thus retaining a constant population size (Annotation 2). Point mutations took place with a probability of 10^{-5} per nucleotide position (Annotation 3). In order to simplify and speed up the computional algorithm the integral parts of the organisms' numbers of surviving progeny were fixed values, *i.e.* independent of statistical fluctuations (Annotation 4).

The fitness f of an organism was chosen to be the sum of two arbitrary values: $f=1+\Delta\gamma$, $\Delta\gamma$ being the *fitness gain* during the experiment. Here, $\Delta\gamma$ was the maximum of the *fitness contributions* $\Delta f(r_a)$ of the individual adaptive subreactions: $\Delta\gamma = \max(\Delta f(r_{a,1}), \Delta f(r_{a,2}), \Delta f(r_{a,3}), \Delta f(r_{a,4}))$, where $r_a \in \{r_{a,1}, r_{a,2}, r_{a,3}, r_{a,4}\}$ signifies an adaptive subreaction, *i.e.* a *partially* or *completely* developed *adaptive reaction*. Thus, the adaptive subreaction r_a have or have not have reached the corresponding target sequence g_a, where $g_a \in \{g_{a,1}, g_{a,2}, g_{a,3}, g_{a,4}\}$. Hence, every approach towards g_a can be regarded as an adaptive subreaction r_a. The corresponding fitness $f(r_a)$ was calculated by the Gauss curve equation $f(r_a)=3\exp_e(-0.001h(g_a)^2)$, $h(g_a)$ being the Hamming distance from the organism to the target sequence g_a. Thus, we simplified actual adaptive landscapes where the higher the adaptation level, *i.e.* specialization level, of organisms in a population, the higher their probability of producing lethal mutations.

After 50,000 generations four medians of the Hamming distances from all organisms to each target sequence were measured. We call these medians *adaptive distances*. At diversity degree 1 all the adaptive reactions occured simultaneously (Figure 1). Thus, four different types of organisms lived in the final population; they had completely developed either the first, the second, the third or fourth adaptive reaction. The corresponding modules were syntheses, because they could be divided into submodules. Furthermore, they were homosyntheses, since they did not contain subnecessities whose adaptive subreactions were mutually exclusive (*i.e.* they could not be divided into incompatible adaptive subreactions). As a reason for the simultaneous coexistence of four homosyntheses (in the experiments with the diversity degree 1) we propose that each homosynthesis was convertible to another through a single point mutation.

At all higher diversity degrees only one adaptive reaction occured, *i.e.* only one homosynthesis developed (Figure 1). As in our virtual subhabitat, this homosynthesis was selected randomly, in a different virtual subhabitat a different homosynthesis might have been selected. As this randomness occured before any completion of adaptive reactions, it is genetic routeing.

Genetic routeing versus genetic drift. Wright called any random disappearance of alleles in small sexual populations genetic drift (WRIGHT 1977). This definition can be extended using our nomenclature: "Genetic drift is a random disappearance of modules whose adaptive reactions have already been completed." (Here, "random disappearance" is a random selection to disappear.) As a complement to genetic drift we define genetic routeing: "Genetic routeing is a random selection of adaptive reactions which have not developed any module yet." (Here, "random selection" is a random selection to survive.) Both randomness mechanisms may principally occur in small and large populations or sexual and asexual species. However, what are their rates and probabilities in the different kinds of populations or species?

Mathematical Multi-Level Evolution Model

KIMURA and OHTA (1968) calculated that genetic drift swiftly occurs in small populations and slowly in large populations. Their diffusion model revealed a linear correlation between the effective population size and the fixation time of an allele, *i.e.* above a certain population size the fixation time became relatively long. The disadvantage of KIMURA and OHTA's (1968) model and other diffusion models is the continuousness of the effective population size, leading to calculation errors in small populations (WATTERSON 1962).

In order to investigate genetic drift and routeing in small and large populations, we have developed a more realistic stochastic model (provided as APPENDIX II) where the population size is discrete. Here, two different module types G and H exist in a habitat of an ecological niche. Consequently, two different kinds of asexual organisms live there – type G and H, or three different kinds of sexual organisms – type GG, GH, and HH.

Given module type G count i and module type H count j in a start population, and the specific fitness values f_G, f_H, f_{GG}, f_{GH}, f_{HH} for the organism kinds G, H, GG, GH, HH, we have calculated module type G extinction probability $x_G(i,j)$ and module type H extinction probability $x_H(i,j)$ and the mean absorption time $y(i,j)$ (defined by GALE 1962), *i.e.* the expected duration of Darwin's struggle for existence. In order to simulate competition among the organisms and a lack of resources, the total number of modules $n_m = i+j$ in the habitat has been kept constant throughout.

Our stochastic model is a generalization of Moran's discrete model (MORAN 1962). We have included continuousness of time and comparability between sexuality and asexuality. Furthermore, we have included distinct expected birth and death rates. This leads to a variance of an organism's number of surviving progeny. Stochastically, this variance is independent of the expectation of an organism's number of surviving progeny, *i.e.* organism-level fitness (see APPENDIX II). Commonly in population genetics, this variance has been subsumed in Wright's effective population size (WRIGHT 1931). We coin a further expression for this variance: *fluctuation*, assuming the species-level fitness to be related to it. We define the fluctuation $v(n)$ of a population containing n organisms as the variance of its increase in organism count per fixed generation time Δt. The fluctuation $v(1)=v$ of a single organism is the fluctuation of a population containing only this organism.

Let v_G, v_H, v_{GG}, v_{GH}, and v_{HH} be the fluctuations of the organism kinds G, H, GG, GH, and HH. Our calculations reveal an almost linear correlation between population size and mean absorption time under the following three conditions (Figure 2): (1st) isoadditive module types G and H (i.e. $\min(f_{GG},f_{HH}) \leq f_{GH} \leq \max(f_{GG},f_{HH})$), ($2^{nd}$) both module types lead to equal fitness, and (3rd) equal module count at the start. Thus, if $f_G=f_{GG}=f_{GH}=f_{HH}=f_H$ and $v_G=v_{GG}=v_{GH}=v_{HH}=v_H$, the discrete gradient $\Delta_{ij}y(i,j)=y(i+1,j+1)-y(i,j)$ is almost constant for any $i=j$ (Figure 3). The limit case $i=j=\infty$ could be regarded as the absolute constancy, *i.e.* linearity, where ∞ is an infinitely large number. For smaller $i=j$ $\Delta_{ij}y(i,j)$ is slightly higher.

The start condition $i=j$ implies that no Darwinian selection process has worked on the module types for a period of time. If a Darwinian selection process had occured, the simultaneous accumulation of the two module types would have been prevented. Thus, all corresponding adaptive reactions must have been developed independently, *i.e.* with little or without competion for resources. An example are isolated habitats. When such habitats are joined together, competition between the corresponding module types is generated, triggering a further evolutionary process: a random disappearance of previously established modules, *i.e.* genetic drift. Therefore, the start condition $i=j$ simulates genetic drift. In this case, our findings about genetic drift are consistent with KIMURA and OHTA's (1968) findings.

For any $i=j$, $f_G=f_{GG}=f_{GH}=f_{HH}=f_H$ and $v_G=v_{GG}=v_{GH}=v_{HH}=v_H$, the gradient $\Delta_{ij}y(i,j)$ is significantly higher in sexual species than in asexual ones (Figures 2 and 3). Thus, genetic drift has a slower rate in sexual species than in asexual ones. This corresponds to the improved ability of sexual species to conserve a variety of different previously established modules. This conceivably fulfills a species-level necessity for a higher degree of genetic polymorphism (BARTON and CHARLESWORTH 1998; BURGER 1999; WAXMAN and PECK 1999).

In sexual and asexual species, the gradient $\Delta_{ij}y(i,j)$ inversely correlates with the fluctuations $v_G=v_{GG}=v_{GH}=v_{HH}=v_H$ (Figure 3). Consequently, if the fluctuations are too large, a random disappearance of modules practicably occurs in a population regardless of size – this is a fast form of genetic drift. However, if the fluctuations are not too large, a random disappearance of modules practicably occurs only in a small population – this is a slow form of genetic drift. Therefore, fluctuations in the net reproduction rates of individual organisms increase the overall rate of genetic drift, *i.e.* they are responsible for this randomness mechanism. Furthermore, we propose that a faster rate of genetic drift can raise the species-level fitness, *i.e.* there is a correlation between fluctuation and species-level fitness.

When would genetic drift occur in our computational habitat? In a further 100 computer experiments the start populations contained 250 organisms per target sequence; all other conditions were identical to the previous experiments. As the integral part of an organism's number of surviving progeny was constant, the fluctuation only resulted from the randomness of its fractional part and from the randomness of point mutations. After 50,000 generations, the experiments with the selected start populations led to the same results as the experiments with the random start populations (Figure 1), demonstrating equilibrium. Thus, at diversity degree 1, there were again four coexistent homosyntheses in the population. Here, genetic drift did not work, since no homosynthesis disappeared. At all higher diversity degrees, only one homosynthesis was found. Here, genetic drift was responsible for the random disappearance of the three other homosyntheses.

What was the rate of genetic drift when homosyntheses disappeared in the selected start populations? In order to answer this question we measured the mean time to this equilibrium at diversity degree 75 in 200 further allopatric speciation experiments: 100 with random start populations and 100 with selected ones. Equilibrium was reached after a number of generations t when the adaptive distances from a population to each target sequence were at their shortest.

In the experiments with the selected start populations, the average t was $A_t = 8882.1$, and in those with the random start populations $A_t = 5059.5$ (Figure 4). In the selected start populations, *i.e.* when all adaptive reactions had already been completed, genetic drift occured. Conversely, in the random start populations, when all adaptive reactions had not started yet to develop, genetic routeing took place. Furthermore, we regard $A_t = 8882.1$ as a relatively long time and $A_t = 5059.5$ as a relatively short time. Thus, genetic drift could be viewed as 'infeasible' in the large populations of 1,000 organism (because it takes too much time), and genetic routeing as 'feasible' (because it does not take to much time). Generally, genetic drift in the allopatric speciation experiments functioned at a slower rate than genetic routeing.

In the allopatric speciation experiments with random start populations a battery of 'new' mutant modules emerged, each initially represented by one organism. Using our stochastic model we have simulated the struggle for existence between a 'new' module and the 'old' modules. This is the simulation of genetic routeing, a randomness mechanism which functions before or during an adaptive reaction, *i.e.* during a Darwinian selection process: If $i=1$, $j=m-1$, $f_G=f_{GG}=f_{GH}=f_{HH}=f_H$, and $v_G=v_{GG}=v_{GH}=v_{HH}=v_H$, the gradient $\Delta_{ij} y(i,j) = y(1,j+2) - y(1,j)$ inversely correlates with the population size, *i.e.* the resulting graphs show concave curves (Figures 5 and 6). Moran's model produced a similar concavity (GALE 1990). This concavity shows that the random disappearance of modules occurs regardless of population size. Thus, the rate of genetic routeing seems to be faster and more independent of population size than the rate of genetic drift (compare Figure 2 with Figure 5).

For any j, $y(1,j)$ is significantly longer in sexual species than in asexual ones (Figure 5). Consequently, a 'new' mutant module has a higher chance of surviving in a sexual species than in an asexual one (BARTON and CHARLESWORTH 1998; BURGER 1999; WAXMAN and PECK 1999). Consequently, genetic routeing functions at a slower rate in sexual species than in asexual ones; this may make diversification in sexual species less likely, *e.g.* allopatric or sympatric speciation. This lower likelihood for speciation processes may increase the degree of genetic polymorphism in sexual species, fulfilling a species-level necessity for flexibility.

When a disadvantage for heterozygotes coincides with genetic drift or routeing, *i.e.* $f_{GH} < f_{GG} = f_{HH}$ for any i and j, $y(i,j)$ decreases (Figures 2 and 5). Such a situation arises when the modules G and H are hypoadditive. This is known as disruptive selection and is widely regarded as being the classical reason for sympatric speciation (MAYNARD SMITH 1966; FELSENSTEIN 1981; KONDRASHOV 1986; DIEHL and BUSH 1989). A hypoadditive relationship promotes a faster rate of genetic routeing or genetic drift. Thus, it makes the diversification of species more likely, or the random disappearance of previously established modules.

When an advantage for heterozygotes coincides with genetic drift or routeing, i.e. $f_{GH} > f_{GG} = f_{HH}$ for any i and j, $y(i,j)$ increases (Figures 2 and 5). Such a situation arises when G and H are hyperadditive. If the population size is large in this case, the gradients $\Delta_{ij} y(i,j)$ and $\Delta_j y(1,j)$ correlate with the population size, creating graphs with convex curves (Figures 2, 5 and 6). This convexity promotes the conservation of modules, since it decreases the feasibility for genetic drift or genetic routeing. Furthermore, we suggest that this module conservation makes unequal crossing over between the hyperadditive modules more likely, which in turn promotes a 'stable' module duplication (SMITH 1976). (Remark: The duplication of isoadditive and hypoadditive modules is 'unstable', since one of the copies could be inactivated by mutation without any major loss of fitness.)

Therefore, sexuality provides two evolutionary mechanisms, which fulfil the species-level necessity for a higher adaptation speed (BARTON and CHARLESWORTH 1998; BURGER 1999; WAXMAN and PECK 1999): a fast diversification of species when different modules are hypoadditive, and an accelerated stable module duplication when they are hyperadditive. The same principle arises when $f_{GG} > f_{HH}$ (Figures 7, 8 and 9): The smaller the fitness f_{GH} of heterozygotes, the shorter the expected duration of the struggle for existence. Here, genetic drift is simulated by $i=j$ (Figure 7), and genetic routeing by $i=1$ (Figure 8) and by $j=1$ (Figure 9). Hence, the smaller the f_{GH}, the faster the rate of genetic drift or genetic routeing. Under comparable conditions, genetic routeing consistently functions at a faster rate than genetic drift (compare Figure 7 with Figures 8 and 9).

Mathematical Multi-Level Evolution Model

We have found an analogous relationship concerning the extinction probability $x_G(i,j)$. In cases of genetic drift, when $f_{GG} > f_{HH}$, the smaller the f_{GH}, the higher the $x_G(i,j)$ (Figure 10). Thus, the smaller the f_{GH}, the larger the range of population sizes where an 'uncertainty principle' occurs. We define 'uncertainty principle' to be a condition of a population where the fittest module G does not inevitably win the struggle for existence, i.e. it also has a probability $0 < x_G(i,j) \leq 1$ of losing. If $i=j$ or $j=1$ occurs, i.e. genetic drift (Figure 10) or genetic routeing after a disadvantageous mutation (Figure 11), the uncertainty principle is restricted to small populations. However, if $i=1$ occurs, i.e. genetic routeing after an advantageous mutation (Figure 12), the uncertainty principle occurs in any population regardless of size.

Sympatric speciation simulation: In the ecological niches of another 200 computer experiments, four necessities prevailed which were able to be fulfilled by four hyperadditive adaptive reactions. The adaptive reations competed for the same genetic locus, thus leading to structural incompatibility. In each ecological niche, the fitness gain $\Delta \gamma$ of an organism was the sum of the fitness contributions $\Delta f(r_a)$ of the individual adaptive subreactions $r_a \in \{r_{a,1}, r_{a,2}, r_{a,3}, r_{a,4}\}$: $\Delta \gamma = \Delta f(r_{a,1}) + \Delta f(r_{a,2}) + \Delta f(r_{a,3}) + \Delta f(r_{a,4}))$. All the other conditions were the same as in the previous computer experiments where 100 experiments were performed with random start populations and 100 with selected start populations. Each selected start population contained 250 organisms per target sequence.

After 50,000 generations, the experiments with the random start populations revealed results similar to those with the selected start populations (Figure 13). In order to describe these results clearly we call *balanced heterosynthesis* a heterosynthesis whose submodules have equal adaptive distances to the corresponding target sequences. Conversely, we call *unbalanced heterosynthesis* a heterosynthesis whose submodules have unequal adaptive distances to the corresponding target sequences.

In a population of either type of experiment, at diversity degree 1 four different homosyntheses developed simultaneouly (being interconvertible by a one-step mutation), from diversity degrees 2 to 28 manifold balanced heterosyntheses between all four necessities developed simultaneously (through manifold combinations of the corresponding adaptive subreactions), and from 29 to 32 manifold balanced heterosyntheses between three necessities developed simultaneously. Here, one necessity was excluded to 'participate' in the heterosyntheses. This necessity was selected randomly, *i.e.* genetic routeing randomly selected the three adaptive reactions which formed the heterosyntheses.

In the experiments with the random start populations, from 33 to 45 (or with the selected start populations from 33 to 44) there were manifold balanced heterosyntheses between two necessities, and from 46 to 63 (or 45 to 62) manifold unbalanced heterosyntheses between two necessities. Again, genetic routeing 'decided' which adaptive reactions 'participated' in the heterosyntheses. Finally, with the random start populations from 64 to 100 (or 63 to 100) there was only one homosynthesis in one population.

Heterosynthesis and homosynthesis are mutually exclusive evolutionary strategies to resolve the dilemma of incompatibility. In an occupied ecological niche, only one homosynthesis can be sustained, unless there is a possible single-step mutation to another homosynthesis, *i.e.* a diversity degree 1. Should different homosyntheses of a higher diversity degree, which fulfil different necessities, occur simultaneously in the same ecological niche, subniches may be occupied. This happens if the populations inhabiting these subniches differ in strategies of resource utilisation or coping with harmful conditions, *i.e.* if they are partially or completely independent in their dynamic regulation of population size. Further transformation or diversification within subniches could ultimately lead to sympatric speciation. In the sympatric speciation experiments with random start populations, genetic routeing was the randomness mechanism underlying the random selection of homosyntheses. In the sympatric speciation experiments with selected start populations, genetic drift was the underlying randomness mechanism.

As in the allopatric speciation computer experiments, we measured the average number of generations A_t to equilibibrium in 200 further experiments at diversity degree 75, *i.e.* 100 experiments with random start populations and 100 with selected ones. In the experiments with the random start populations, $A_t=16,763.1$, and in those with the selected start populations, $A_t=18,456.2$ (Figure 14). Here, both values could be regarded as impracticable time, as they were much longer than those in the allopatric speciation experiments (Figure 4). Consequently, both genetic drift and genetic routeing functioned at a slower rate in sympatric speciation than in allopatric speciation. This corresponds to our calculations with hyperadditive modules in sexual species, indicating a lengthy struggle for existence (Figures 2 and 5). In the sympatric speciation experiments, genetic drift also seemed to be slightly 'slower' than genetic routeing, but this difference was not as clear as the one in the allopatric speciation experiments.

Genomic incompatibility (*i.e.* incompatibility of hyperadditive adaptive reactions) might lead to a 'slow' sympatric speciation process. We propose such a scenario took place during the Cambrian explosion (OHNO 1997) when bursts of necessities might have compensated for this 'slowness'. After the Cambrian explosion, however, its legacy constraints (OHNO

1997) might have resulted in a reduction in the frequency of necessities. This would be consistent with Mayr's claim that sympatric speciation is unlikely in actual ecological niches (MAYR 1963). Nevertheless, another sympatric speciation process, which might have been independent of a Cambrian necessity burst, was found to occur under certain conditions such as disruptive selection (MAYNARD SMITH 1966; FELSENSTEIN 1981; KONDRASHOV 1986; DIEHL and BUSH 1989), *i.e* a hypoadditive relationship between adaptive reactions. Corresponding to our calculations with hypoadditive adaptive reactions in sexual species, which have indicated a brief struggle for existence (Figures 2 and 5), this sympatric speciation process could be regarded as 'fast'.

Measuring adaptive success. As a diagnostic means to differentiate between homosynthesis and heterosynthesis we propose a pA value (point of adaptation):

$$\text{pA} = -\sum_{u=1}^{l} \log_{10} \sum_{n(u)}^{k} \binom{k}{n(u)} \times 0.25^{n(u)} \times 0.75^{k-n(u)}$$

Here, k is the number of sequences in the population, l the length of the sequences, u each nucleotide position, $n(u)$ the number of the most frequent nucleotide at position u in the population; pA is the negative common logarithm of the probability that a nucleotide at every position u occurs at least $n(u)$ times.

When $k=1,000$ and $l=100$, the pA value of a randomly generated population is *approx.* 95. The pA value of a population exclusively comprising one type of sequence would be *approx.* 30,765. Thus, the lower the degree of genetic polymorphism, the higher the pA value. This corresponds to Fischer's fundamental theorem (FISHER 1930) about the inverse correlation between the 'average fitness' of a population, *i.e.* its adaptation level, and the 'genetic variance'. The pA value in all our computer experiments was measured after 50,000 generations (Figure 15).

Experiments with isoadditive adaptive reactions: When only one homosynthesis evolved in a population, pA≈30,765. When four homosyntheses evolved simultaneously from the random start population, pA≈30,500 (or from the selected start population, pA≈30,492).

Experiments with hyperadditive adaptive reactions: When only one homosynthesis evolved from the random start populations the average pA value was A_{pA}≈30,726 (or A_{pA}≈30,721) and its standard deviation S_{pA}≈91 (or S_{pA}≈108). When four homosyntheses evolved simultaneously from the random start population, pA≈30,502 (or from the selected start population, pA≈30,484). When manifold heterosyntheses evolved from the random start populations, A_{pA}≈27,334 (or from the selected start populations, A_{pA}≈26,975) and S_{pA}≈2,432 (or S_{pA}≈2,676).

Hence, heterosynthesis on average reduces the pA value, thus increasing the degree of genetic polymorphism.

DISCUSSION

How and when did heterosyntheses evolve, *e.g.* the bifunctional cytoplasmic aconitase (HENTZE and KUEHN 1996)? Our computer experiments with asexual species and further stochastic analysis with sexual and asexual species revealed an increase in the expected duration of the struggle for existence in cases of genomic incompatibility, *i.e.* when surplus necessities with hyperadditive adaptive reactions compete for scarce genetic loci. The degree of this increase might be a criterion limiting the likelihood of heterosyntheses evolving. When the expected duration is too long, module duplication occurs and the likelihood of heterosyntheses decreases. Thus, the expected duration has to be sufficiently short in order for heterosyntheses to evolve.

Another criterion for heterosyntheses is the quantity of necessities triggered by one situational change. When the number of necessities is sufficiently large, a corresponding number of module duplications, *i.e.* time (if we assume a more or less constant duplication rate), is required for all these necessities to be satisfied. This prolongs the duration of genomic incompatibility and increases the likelihood of its being resolved by heterosynthesis. We propose that such a scenario happened during the Cambrian explosion. We suggest that most heterosyntheses such as the bifunctional cytoplasmic aconitase evolved here, due to bursts of necessities. During this geological epoch 500 to 550 million years ago, radical changes in body complexity of triblastians were universally established (OHNO 1997). This was accompanied by the progressive expansion of the Hox gene cluster due to multiple duplications (DEROSA et al. 1999). The Cambrian necessity bursts increased the likelihood of duplicated module copies to become hyperadditive, thus decreasing the rate of mutational inactivation of one of the copies.

We propose that there is a higher probability of stable module duplications in sexual species than in asexual species. This is supported by our stochastic analysis, where the expected duration of the struggle for existence is calculated to be significantly longer in sexual species than in asexual ones (Figures 2, 3, 5, 7, 8 and 9). We suggest that this longer expected duration increases the probability of unequal crossing over (SMITH 1976) between different module 'alleles' of a single genetic locus. Here, the linear incorporation of a second module 'allele' into a chromosome which carries a first one, would create a second genetic locus. If the different modules are hyperadditive (assuming that equal modules usually are isoadditive) both genetic loci are stable. A stable module duplication can be regarded as an adaptive reaction which is likely to enhance the feasibility of other adaptive reactions, thus triggering further evolution. Moreover, an unstable module duplicate could also trigger further evolution, however, depending on its rate of decay.

Stable and unstable module duplicates were the foundation for the cascade of feasible adaptive reactions during the Cambrian explosion. However, their evolution was dependent on bursts of necessities, without which the unstable duplicates would have decayed to pseudogenes (MIGHELL et al. 2000). What situational change triggered such bursts of necessities in early triblastians, i.e. some sexual species, and consequently the Cambrian explosion? We propose that it was early multicellularity together with endosymbiosis of mitochondria (GRAY et al. 1999). Intracellular reproduction of these organelles amplified citric acid cycle modules, thus increasing the oxidative metabolism rate. This situational change triggered bursts of necessities, e.g. for differentiation into multiple anatomical organs. The corresponding adaptive reactions were different combinations of diversified modules, resulting in diversification of biological functions, i.e. triblastian species (MUELLER and MUELLER 1998).

Early triblastians utilized duplicated modules as diversifying 'building blocks' to create 'multicellular complexity'. (With 'multicellular complexity' we refer to a high degree of functional interdependency in multicellular organisms such as vertebrates.) This hierarchical tree of 'old' and 'young' modules is genetically expressed in different patterns, each specific for one anatomical cell type. Multicellular complexity could only have developed when all the 'building blocks' of a species matched at every evolutionary stage. This prerequisite may account for Haeckel's observation that a hierarchical ontogeny recapitulates phylogeny (GOULD 1977). Furthermore, multicellular complexity reduces the set of feasible adaptive reactions. One hierarchical module tree can oust competing module trees. This competition is literally Dawkin's 'selfishness' of genes, *i.e.* genes stubbornly replicate and diversify (DAWKINS 1976), regardless of a possible reduction in adaptability, *i.e.* species-level fitness.

Conversely, in 'lower' clades such as procaryotes and unicellular eucaryotes, basic modules were less frequently duplicated, and thus less frequently utilized as 'building blocks'. A lack of multicellular complexity would enlarge the set of feasible adaptive reactions, leading to unicellular complexity. (With 'unicellular complexity' we refer to a high degree of functional interdependency in unicellular organisms such as bacteria.) Lenski's artificial life system showed the superior mutation tolerance of species with unicellular complexity (LENSKI *et al.* 1999). Modules in such 'lower' clades seem to be less 'selfish', since they less frequently oust other modules. Therefore, we propose a generally higher rate of adaptability, *i.e.* species-level fitness, in 'lower' clades than in triblastians.

Why did the situational change triggering the Cambrian explosion in triblastians, *i.e.* the combination of multicellularity, endosymbiosis and sexuality, not occur in species of 'lower' clades instead? Was the occurrence of this situational change in the predecessors of early triblastians due to randomness or adaptation? The mathematical model of our 'logical evolutionary language' (APPENDIX I) may be helpful in answering this question in future research.

Like the hierarchy of a phylogenic tree of clades, the situational change which caused the Cambrian explosion can be subdivided into infinitely many levels of situational subchanges. At each level one or more subnecessities are sparked off, which in turn can be subdivided further. This is consistant with fractal self-similarity (MANDELBROT 1988; EIGEN *et al.* 1993; EIGEN *et al.* 1996). Each subnecessity may be fulfilled by one or more adaptive subreactions, which lead to new situational subchanges.

A situational subchange may trigger an infinite number of subnecessities. There can be an infinite number of adaptive subreactions, feasible and unfeasible, relating to these subnecessities. However, what happens if the supply of feasible adaptive subreactions is finite and eventually becomes exhausted? It seems plausible that such a finiteness is a consequence of the finite nucleic acid structure. Darwin could not have known this finiteness of phenotypic variation possibilities (DARWIN 1859, *e.g. pg.* 115): *"But Natural Selection, as we shall hereafter see, is a power incessantly ready for action, and is as immeasurably superior to man's feeble efforts, as the works of Nature are to those of Art."* Darwin's dualism between *"Nature"* and *"Art"* corresponds to Descartes's rationalism, where *"Nature"* as God's work was associated with infinite substance, and *"Art"* as a human creative skill was associated with finite substance (DESCARTES and LAFLEUR 1956).

Darwin's phenotypic variation possibilities can be related to a tree of feasible adaptive reactions. However, he assumed these variation possibilities to be infinite. Therefore, we liken Darwin's infinite number of phenotypic variation possibilities to an infinite tree of satisfiable and unsatisfiable necessities. Furthermore, we propose a finite tree comprising corresponding feasible adaptive reactions. Using up a finite supply of feasible adaptive reactions in an ecological niche halts evolution within it. Gould and Eldredge called this stasis punctuated equilibrium (GOULD and ELDREDGE 1977; GOULD and ELDREDGE 1993). Further evolution can only restart if new situational changes trigger necessities. Nevertheless, the low adaptability of organisms with multicellular complexity usually makes necessities unsatisfiable. In this case, such clades become extinct, *e.g.* highly specialized clades such as many arthropods from the Burgess Shale, *e.g. Yohoia tenuis* (WHITTINGTON 1974).

The punctuated equilibrium concept implies a much faster evolutionary process during the 'punctuated' periods than Darwin could presumably expect (DARWIN 1859). This corresponds to an accelerated evolutionary process in Eigen and Schuster's quasispecies distribution model (EIGEN 1971; EIGEN *et al.* 1989; EIGEN *et al.* 1996). Lenski found the phenomenon of punctuated equilibrium in *E.coli* (ELENA *et al.*). This phenomenon may have been instrumental in the fast and possibly random diversification of Darwin's finches on the Galapagos islands (GRANT 1986).

We investigated such a possible randomness under computational and reproducible conditions. Conforming with Kimura's neutral theory of selection (KIMURA 1991) and Wright's shifting balance theory (WRIGHT 1982b), our experimental 'quasispecies' was subjected to isoadditive adaptive reactions which were 'neutral' to each other, *i.e.* they led to equal fitness. Genetic routeing selected just one to develop. This suggests that in actual ecological niches only one adaptive reaction would be detectable, unless there were more geographical habitats with the same necessities.

Perhaps genetic routeing was instrumental in the development of genetic code. Its universality could indicate that it must have originated in one habitat. The ancestral population of all contemporary living species may have existed there. This supports the theory that life on earth originated from a common ancestral population. Furthermore, genetic routeing may be responsible for aconitase being the iron concentration sensing regulator in ferritin metabolism, and not any other iron sulphur cluster containing enzyme.

Genetic routeing could also have been responsible for the fact that on different Galapagos islands Darwin's finches evolved in different directions. These geologically young vulcanic islands had probably been inhabited by a common ancestral species of finches from the mainland (GRANT 1986). At that time, each island might have been a habitat with a same common set of necessities.

LENSKI and TRAVISANO (1994) investigated 12 habitats with the same necessities, *i.e.* experimental conditions, occupied by the same subpopulations of *E. coli*. The subpopulations randomly diverged in mean cell size and reproduction rate (LENSKI and TRAVISANO 1994). The mean time to equilibrium was short, *i.e.* 2000 generations, and population sizes were large, *i.e.* fluctuating between 5×10^6 to 5×10^8 bacteria. Following our results we propose genetic routeing to be the randomness mechanism in these experiments.

We conclude, that allopatric speciation in procaryotes may be due to random diversification in different directions, *i.e.* towards mutually exclusive adaptive peaks. Furthermore, procaryotic allopatric speciation does not seem to be dependent on small population size, *i.e.* the founder effect (WRIGHT 1977).

LITERATURE CITED

BARTON, N. H., and B. CHARLESWORTH, 1998 Why sex and recombination? *Science* **281:** 1986-1990.

BEGON, M., MORTIMER, M., and D. J. THOMPSON, 1996 *Population Ecology*. Blackwell Scientific Publications, Oxford.

BRÉMAUD, P., 1999 *Markov chains: Gibbs fields, Monte Carlo simulation, and queues*. Springer, New York.

BRODEUR, G. M., and M. D. HOGARTY, 1998 Gene Amplification in Human Cancers: Biological and Clinical Significance, pp. 161-172 in *The Genetic Basis of Human Cancer*, edited by B. VOGELSTEIN, and K. W. KINZLER. Mc Graw-Hill.

BROSIUS, F., and S. J. GOULD, 1992 On "genomenclature": a comprehensive (and respectful) taxonomy for pseudogenes and other "junk DNA". *Proc. Natl. Acad. Sci. USA* **89:** 10706-10710.

BURGER, R., 1999 Evolution of genetic variability and the advantage of sex and recombination in changing environments. *Genetics* **153:** 1055-1069.

DARWIN, C., 1859 *The Origin of Species*. Penguin Books, London.

DAWKINS, R., 1976 *The selfish gene*. Oxford Univ. Press, Oxford.

DAWKINS, R., 1988 The one true tree of life, pp. 255-284 in *The Blind Watchmaker*. Penguin Books, London.

DEROSA, R., GRENIER, J. K., ANDREEVA, T., COOK, C. E., ADOUTTE, A., AKAM, M., CARROLL, S. B., and G. BALAVOINE, 1999

Hox genes in brachiopods and priapulids and protostome evolution. *Nature* **399**: 772-776.

DESCARTES, R. and L. J. LAFLEUR, 1956 *Discourse on Method*. Prentice Hall.

DIEHL, S. R. and G. L. BUSH, 1989 The role of habitat preference in adaptation and speciation, pp. 345-365 in *Speciation and Its Consequences*, edited by D. OTTE and J. A. ENDLER. Sinauer Associates, Sunderland, MA.

EBBINGHAUS, H.F., FLUM, J. AND THOMAS, W., 1994 *Mathematical Logic*. Springer New York

EIGEN, M., 1971 Selforganization of Matter and the Evolution of Biological Macromolecules. *Naturwissenschaften* **58**: 465-523.

EIGEN, M., MCCASKILL, J. and P. SCHUSTER, 1989 *Adv. Chem. Phys.* **75**: 149-263.

EIGEN, M., WINKLER, R., and R. KIMBER, 1993 *Laws of the Game: How the Principles of Nature Govern Chance*. Princeton Univ. Press.

EIGEN, M., WINKLER-OSWATITSCH R., and P. WOOLLEY, 1996 *Steps Towards Life: A Perspective on Evolution*. Oxford Univ. Press.

ELENA, S. F., COOPER, V. S. and R. E. LENSKI, 1996 Punctuated evolution caused by selection of rare beneficial mutations. *Science* **272**: 1802-1804.

FELSENSTEIN, J., 1981 Skepticism towards Santa Rosalia, or why are there so few kinds of animals? *Evolution* **35**: 124-138.

FISHER, R. A., 1930 *The general theory of natural selection*. Clarendon Press, Oxford.

GALE, J. S., 1990 Mean sojourn, absorption and fixation times, pp. 213-276 in *Theoretical Population Genetics*. Unwin Hyman, London.

GEHRING, N. H., HENTZE, M. W., and K. PANTOPOULOS, 1999 Inactivation of both

RNA binding and aconitase activities of iron regulatory protein-1 by quinone-induced oxidative stress. *J. Biol. Chem.* **274:** 6219-6225.

GOULD, S. J., 1977 *Ontogeny and Phylogeny.* Harvard University Press, Cambridge, MA.

GOULD, S. J., and N. ELDREDGE, 1977 *Paleobiology* **3:** 115-151.

GOULD, S. J., and N. ELDREDGE, 1993 Punctuated equilibrium comes of age. *Nature* **366:** 223-227.

GOULD, S. J., and E. A. LLOYD, 1999 Individuality and adaptation across levels of selection: How shall we name and generalize the unit of Darwinism? *Proc. Natl. Acad. Sci. USA* **96:** 11904-11909.

GOULD, S. J., and E. S. VRBA, 1982 *Paleobiology* **8:** 4-15.

GRANT, P. R., 1986 *Ecology and Evolution of Darwin's Finches.* Princeton Univ. Press, Princeton.

GRAY, M. W., BURGER, G., and B. F. LANG, 1999 Mitochondrial evolution. *Science* **283:** 1476-1481.

HAMMING, R. W., 1980 *Coding and Information Theory.* Englewood Cliffs, Prentice Hall.

HENTZE, M. W., and L. C. KUEHN, 1996 Molecular control of vertebrate iron metabolism: mRNA-based regulatory circuits operated by iron, nitric oxide, and oxidative stress. *Proc. Natl. Acad. Sci. USA* **93:** 8175-8182.

HENTZE, M. W., ROUAULT, T. A., HARFORD, J.B., and R. D. KLAUSNER, 1989 Oxidation-reduction and the molecular mechanism of a regulatory RNA-protein interaction. *Science* **244:** 357-359.

HUTCHINSON, G. E., 1957 Concluding remarks. *Cold Spring Harbor Symp. Quant. Biol.* **22:** 415-427.

HUYNEN, M. A., DANDEKAR, T., and P. BORK, 1999 Variation and evolution of the citric-acid cycle: a genomic perspective. *Trends in Microbiology* **7**: 281-291.

KIMURA, M., 1991 Recent development of the neutral theory viewed from the Wrightian tradition of theoretical population genetics. *Proc. Natl. Acad. Sci. USA* **88**: 5969-5973.

KIMURA, M., and T. OHTA, 1968 The Average Number of Generations Until Fixation of a Mutant Gene in a Finite Population. *Genetics* **61**: 763-771.

KNIGHT, R. D., FREELAND, S. J., and L. F. LANDWEBER, 1999 Selection, history and chemistry: the three faces of the genetic code. *Trends Biochem. Sci.* **24**: 241-247.

KONDRASHOV, A. S. 1986 Multilocus model of sympatric speciation. III. Computer simulations. *Theor. Pop. Biol.* **29**: 1-15.

KREBS, H. A., 1970 The history of the tricarboxylic acid cycle. *Perspect. Biol. Med.* **14**: 154-170.

LAMARCK, J. B., 1809 *Philosophie Zoologique*. Paris.

LENSKI, R. E. and M. TRAVISANO, 1994 Dynamics of adaptation and diversification: a 10,000-generation experiment with bacterial populations. *Proc. Natl. Acad. Sci. USA* **91**: 6808-6814.

LENSKI, R. E., OFRIA, C., COLLIER, T. C. and C. ADAMI, 1999 Genome complexity, robustness and genetic interactions in digital organisms. *Nature* **400**: 661-664.

LEWIN, B., 1997 Condensing viral genomes into their coats, pp. 744-747 in *Genes VI*. Oxford University Press, New York.

MANDELBROT, B. B., 1988 *The Fractal Geometry of Nature*. W H Freeman & Co.

MAYNARD SMITH, J., 1966 Sympatric speciation. *American Naturalist* **100**: 637-650.

MAYR, E., 1942, *Systematics and the Origin of Species*.

Columbia University Press, New York.

MAYR, E., 1963, *Animal Species and Evolution*. Harvard University Press, Cambridge, MA.

MAYR, E., 1992 Darwin's Principle of Divergence. *J. Hist. Biol.* **25**: 343-359.

MIGHELL, A. J., SMITH, N. R., ROBINSON, P. A., and A. F. MARKHAM, 2000 Vertebrate pseudogenes. *FEBS Lett.* **468**: 109-114.

MORAN, P. A. P., 1962 The survival of a single mutant, pp. 104-121 in *The statistical Processes of Evolutionary Theory*. Oxford University Press, London.

MUCKENTHALER, M., GUNKEL, N., FRISHMAN, D., CYRKLAFF, A., TOMANCAK, P., and M. W. HENTZE, 1998 Iron-regulatory protein-1 (IRP-1) is highly conserved in two invertebrate species – characterization of IRP-1 homologues in *Drosophila melanogaster* and *Caenorhabditis elegans*. *Eur. J. Biochem.* **254**: 230-237.

MUELLER, W. E. G., and I. M. MUELLER, 1998 Transition from Protozoa to Metazoa: An Experimental Approach. *Progress in Molecular and Subcellular Biology* **19**: 1-22.

NARAHARI, J. MA, R., WANG, M., and W. E. WALDEN, 2000 The aconitase function of iron regulatory protein 1. Genetic studies in yeast implicate its role in iron-mediated redox regulation. *J. Biol. Chem.* **275**: 16227-16234.

OHNO, S., 1997 The Reason for as Well as the Consequence of the Cambrian Explosion in Animal Evolution. *Mol. Evol.* **44 (Suppl. 1)**: 23-27.

ORGEL, L. E., 1998 The origin of life – a review of facts and speculations. *Trends. Biochem. Sci.* **23**: 491-495.

PARASKEVA, E., and M. W. HENTZE, 1996 Iron-sulphur clusters as genetic regulatory switches: the bifunctional iron regulatory protein-1. *FEBS Lett.* **389**: 40-43.

PRESS, W. H., TEUKOLSKY, S. A., VETTERLING, W. T., and B. P. FLANNERY, 1992 An Even

Quicker Generator, pp. 284-285 in *Numerical Recipes in C.* Cambridge University Press.

RADMAN, M., 1999 Mutation: Enzymes of evolutionary change. *Nature* **401**: 866-869.

ROUAULT, T. A., and R. D. KLAUSNER, 1997 Regulation of iron metabolism in eukaryotes. *Curr. Top. Cell. Regul.* **35**: 1-19.

SERENO, P. C., 1999 The Evolution of Dinosaurs. *Science* **284**: 2137-2147.

SMITH, G. P., 1976 Evolution of repeated DNA sequences by unequal crossover. *Science* **191,** 528-535.

WATTERSON, G. A., 1962 Some theoretical aspects of diffusion theory in population genetics. *Ann. Math. Stat.* **33:** 939-957.

WAXMAN, D., and J. R. PECK, 1999 Sex and adaptation in a changing environment. *Genetics* **153,** 1041-1053.

WHITTINGTON, H.B., 1974 Yohoia Walcott and Plenocaris n. gen., arthropods from the Burgess Shale, Middle Cambrian, British Columbia. *Geological Survey of Canada Bulletin* **231**:1-21.

WRIGHT, S., 1931, Evolution in Mendelian populations. *Genetics* **16:** 97-159.

WRIGHT, S., 1932 The roles of mutation, inbreeding, crossbreeding and selection in evolution. *Proc. 6th Int. Congr. Gen.* **1:** 356-366.

WRIGHT, S., 1977 Experimental Results and Evolutionary Deductions, pp. 443-473 in *Evolution and the Genetics of Populations, Vol. 3.* University of Chicago Press, Chicago.

WRIGHT, S., 1982a Character change, speciation, and the higher taxa. *Evolution* **36**: 427-443.

WRIGHT, S., 1982b The shifting balance theory and macroevolution. *Ann. Rev. Genet.* **16:** 1-19.

WRIGHT, S., 1988 Surfaces of Selective Value Revisited. *Am. Nat.* **131:** 115-123.

ANNOTATIONS

Annotation 1: In our random DNA generation, the linear congruential random number generator of the Borland® C^{++} *ver.* 5.01 compiler was used.

Annotation 2: The fitness of each organism was subtracted by a generation-specific value calculated by the logistic difference equation according to BEGON *et al.* 1996. Organisms with a fitness smaller than or equal to -1 died immediately. When the total organism count was larger than 1,000 the quick random number generator according to PRESS *et al.* 1992 selected further organisms to die.

Annotation 3: The quick random number generator according to PRESS *et al.* 1992 determined whether a mutation in each organism occured and at which DNA position.

Annotation 4: During a fixed generation time in our C^{++} progam, the algorithm was governed by the following main steps: 1. The organisms' numbers of surviving progeny equaled the organisms' individual fitness values. 2. The individual fitness values of all organisms were calculated. 3. The quick random number generator according to Press *et al.* 1992 selected floating point random numbers between zero and one, where each number was associated with an organism. 4. When the random number associated with an organism was smaller than or equal to the fractional part of the organism's fitness, the integral part of the organism's fitness was augmented by one. When the random number associated with an organism was larger than the fractional part of the organism's fitness, the integral part of the organism's fitness remained constant. 5. The integral part of an organism's fitness was the organism's number of surviving progeny.

Mathematical Multi-Level Evolution Model

APPENDIX I: FORMAL EVOLUTIONARY LANGUAGE

Axiom. An *evolutionary space* is a quintuple $(\Omega_A; \Omega_I; \Omega_C; \Omega_H; \mathbb{R}_0^+)$ of five sets, each of which containing a different kind of atomic quality.

a) Ω_A is the set of all *atomic phenotypical attributes*.

b) Ω_I is the set of all *atomic genotypical instructions*.

c) Ω_C is the set of all *atomic external conditions*.

d) Ω_G is the set of all *atomic geographical regions*.

e) $\mathbb{R}_0^+$ is the set of all *time points*.

Definition 1.

a) A *phenotypical attribute* is a set of subsets of Ω_A. A indicates the set of all possible phenotypical attributes.

b) A *genotypical instruction* is a set of subsets of Ω_I. I indicates the set of all possible genotypical instructions.

c) An *external condition* is a set of subsets of Ω_C. C indicates the set of all possible external conditions.

d) A *geographical region* is a set of subsets of Ω_G. G indicates the set of all possible geographical regions.

e) A *generation* is a subset of $\mathbb{R}_0^+$. Let T be the power set of $\mathbb{R}_0^+$, indicating the set of all possible generations.

Mathematical Multi-Level Evolution Model

Definition 2.

a) A *phenotype* is a subset of A. The power set of A is P, indicating the set of all possible phenotypes.

b) A *genotype* is a subset of I. The power set of I is G, indicating the set of all possible genotypes.

c) An *ecological niche* is a subset of C. The power set of C is N, indicating the set of all possible ecological niches.

d) A *habitat* is a subset of G. The power set of G is H, indicating the set of all possible habitats.

e) A *change* u is either an attribute or an external condition. Thus, $u \in A \cup C$.

Definition 3. Let $p \in P$ be a phenotype, $g \in G$ be a genotype, $n \in N$ be an ecological niche.

a) An *organism* o is a set $o := p \cup g$. The set of all possible organisms is O, being the power set of $A \cup I$.

b) A *situation* s is a set $s := o \cup n$. The set of all possible situations is S, being the power set of $A \cup I \cup C$.

Mathematical Multi-Level Evolution Model

Definition 4.

Let s be a situation, o be an organism, n be an ecological niche, t be a generation, and h be a habitat.

a) '...Ξ...' is a binary relation meaning '(situation)...*exists at/in* ...(generation;habitat)', so that $s \; \Xi \; (t;h)$ is a logic formula, where $(t;h)$ is an ordered pair.

b) A *population* $l_{n,h}(t)$ is a set $l_{n,h}(t) := \{o : o \cup n \; \Xi \; (t;h)\}$.

c) Let $\tau \subset T$ be a set of generations, $t \in \tau$ be a generation, $l_{n,h}(t) := \{o_1, o_2, o_3, \ldots o_k\}$ be a population, and $\psi : l_{n,h}(t) \to k$ be a function which indicates the number of elements in $l_{n,h}(t)$. A *capacity* $\chi(\tau) \in \mathbb{N} \cup \{\infty\}$ of the population $l_{n,h}(t)$ at every $t \in \tau$ is a number, where $(\forall t \in \tau \; \psi(l_{n,h}(t)) \leq \chi(\tau)) \wedge (\exists t \in \tau \; \psi(l_{n,h}(t)) = \chi(\tau))$.

Definition 5. Let $g_i \in G$ be a genotype, $p_i \in P$ be a phenotype, and $o_i := \{g_i, p_i\}$ with $i \in \mathbb{N} \cup \{\infty\}$ be an organism. Then, a *species* ξ of a population l is defined as

$$\xi := \left\{ \bigcap_{g_i \in o_i \in l} g_i, \pi \cup \bigcap_{p_i \in o_i \in l} p_i \right\}.$$

Definition 6.

a) A *fitness* $f(o,n)$ of an organism o in a niche n is a real number.

b) A *fluctuation* $y(o,n)$ of an organism o in a niche n is a positive real number.

Mathematical Multi-Level Evolution Model

Definition 7. Let a_1, a_2 be two phenotypical attributes, o be an organism, and n be an ecological niche.

a) a_1 and a_2 are *hyperadditive* if and only if $f(o \cup \{a_1,a_2\},n) > \max(f(o \cup \{a_1\},n); f(o \cup \{a_2\},n))$.

b) a_1 and a_2 are *isoadditive* if and only if

$\min(f(o \cup \{a_1\},n); f(o \cup \{a_2\},n)) \leq f(o \cup \{a_1,a_2\},n) \leq \max(f(o \cup \{a_1\},n); f(o \cup \{a_2\},n))$.

c) a_1 and a_2 are *hypoadditive* if and only if $f(o \cup \{a_1,a_2\},n) < \min(f(o \cup \{a_1\},n); f(o \cup \{a_2\},n))$.

Definition 8. Let $t_1 < t_2 < t_3$ be three time points, $a \in A$ be a phenotypical attribute, $i \in I$ be a genotypical instruction, $u \in A \cup C$ a change, $h \in H$ be a habitat, o be an organism, and n be an ecological niche.

a) A *situational change* $\sigma_u(t_1,t_2)$ through the change $u \neq \emptyset$ is a logic formula

$\sigma_u(t_1,t_2) := \exists o \in O \; \exists i \in I \; (o \cup n \backslash \{u,i\} \; \Xi \; (t_1;h)) \wedge (o \cup n \cup \{u,i\} \; \Xi \; (t_2;h))$.

b) A *necessity* $v_a(t_2)$ of a phenotypical attribute $a \neq \emptyset$ is a logic formula

$v_a(t_2) := \exists o \in O \; (o \cup n \backslash \{a\} \; \Xi \; (t_2;h)) \wedge f(o \cup \{a\},n) > f(o \backslash \{a\},n)$.

c) An *adaptive reaction* $\rho_i(t_2,t_3)$ of a genotypical instruction $i \neq \emptyset$ is a logic formula

$\rho_i(t_2,t_3) := \exists o \in O \; \exists a \in A \; (o \cup n \backslash \{i,a\} \; \Xi \; (t_2;h)) \wedge (o \cup n \cup \{i,a\} \; \Xi \; (t_3;h))$.

Mathematical Multi-Level Evolution Model

Definition 9. Let σ_1, σ_2 be situational changes, ν_1, ν_2 be necessities, ρ_1, ρ_2 be adaptive reactions, $t_{1,1} < t_{1,2} < t_{1,3}, t_{2,1} < t_{2,2} < t_{2,3}$ be six time points, o be an organism, n be an ecological niche, h_1, h_2 be habitats, $u \neq \emptyset$ be a change, $a \in A \backslash \emptyset, \alpha \in A$ be phenotypical attributes, and $i \in I \backslash \emptyset, \iota \in I$ be genotypical instructions.

a) σ_1 and σ_2 are *analogous* if and only if

$$\forall j \in \{1,2\}\ \sigma_j \Leftrightarrow \exists o \in O\ \exists \iota \in I\ (o \cup n \backslash \{u, \iota\} \ \Xi\ (t_{j,1}; h_j)) \wedge (o \cup n \cup \{u, \iota\}\ \Xi\ (t_{j,2}; h_j)).$$

b) ν_1 and ν_2 are *analogous* if and only if

$$\forall j \in \{1,2\}\ \nu_j \Leftrightarrow \exists o \in O\ (o \cup n \backslash \{a\}\ \Xi\ (t_{j,2}; h_j)) \wedge f(o \cup \{a\}, n) > f(o \backslash \{a\}, n).$$

c) ρ_1 and ρ_2 are *analogous* if and only if

$$\forall j \in \{1,2\}\ \rho_j \Leftrightarrow \exists o \in O\ \exists \alpha \in A\ (o \cup n \backslash \{i, \alpha\}\ \Xi\ (t_{j,2}; h_j)) \wedge (o \cup n \cup \{i, \alpha\}\ \Xi\ (t_{j,3}; h_j)).$$

Definition 10. Let $\sigma, \sigma_1, \sigma_2$ be situational changes, ν, ν_1, ν_2 be necessities, and ρ, ρ_1, ρ_2 be adaptive reactions.

a) σ_1 and σ_2 are *connected by AND* (or *OR* or *XOR*) if and only if $\exists \sigma\ \sigma_1 \wedge \sigma_2 \Leftrightarrow \sigma$ (or $\sigma_1 \vee \sigma_2 \Leftrightarrow \sigma$ or $\sigma_1 \oplus \sigma_2 \Leftrightarrow \sigma$).

b) ν_1 and ν_2 are *connected by AND* (or *OR* or *XOR*) if and only if $\exists \nu\ \nu_1 \wedge \nu_2 \Leftrightarrow \nu$ (or $\nu_1 \vee \nu_2 \Leftrightarrow \nu$ or $\nu_1 \oplus \nu_2 \Leftrightarrow \nu$).

c) ρ_1 and ρ_2 are *connected by AND* (or *OR* or *XOR*) if and only if $\exists \rho\ \rho_1 \wedge \rho_2 \Leftrightarrow \rho$ (or $\rho_1 \vee \rho_2 \Leftrightarrow \rho$ or $\rho_1 \oplus \rho_2 \Leftrightarrow \rho$).

Mathematical Multi-Level Evolution Model

Definition 11. Let $\rho(t_1,t_2)$ be an adaptive reaction, $\nu(t_1)$ be a necessity, and $t_1<t_2$ be two time points.

a) $\rho(t_1,t_2)$ is *nonrandom* if and only if $\exists \nu(t_1)\ \rho(t_1,t_2) \Rightarrow \nu(t_1)$.

b) $\rho(t_1,t_2)$ is *random* if and only if $\neg(\exists \nu(t_1)\ \rho(t_1,t_2) \Rightarrow \nu(t_1))$.

Definition 12. Let $\sigma(t_1,t_2)$ be a situational change, $\nu(t_2)$ be a necessity, $\rho(t_2,t_3)$ be an adaptive reaction, and $t_1<t_2<t_3$ be three time points. Then, a *module* μ is a logic formula

$$\mu := \sigma(t_1,t_2) \Rightarrow \nu(t_2) \Leftarrow \rho(t_2,t_3).$$

Definition 13. Let n_1, n_2 be ecological niches, h_1, h_2 habitats, σ_1, σ_2 be situational changes, ν_1, ν_2 be necessities, ρ_1, ρ_2 be adaptive reactions, and μ_1, μ_2 be modules.

a) n_1 is the *ecological subniche* of n_2 if and only if $n_1 \supseteq n_2$. Then, n_2 is the *ecological superniche* of n_1.

b) h_1 is the *subhabitat* of h_2 if and only if $h_1 \supseteq h_2$. Then, h_2 is the *superhabitat* of h_1.

c) σ_1 is the *situational subchange* of σ_2 if and only if $\sigma_2 \Rightarrow \sigma_1$. Then, σ_2 is the *situational superchange* of σ_1.

d) ν_1 is the *subnecessity* of ν_2 if and only if $\nu_2 \Rightarrow \nu_1$. Then, ν_2 is the *supernecessity* of ν_1.

e) ρ_1 is the *adaptive subreaction* of ρ_2 if and only if $\rho_2 \Rightarrow \rho_1$. Then, ρ_2 is the *adaptive superreaction* of σ_1.

f) μ_1 is the *submodule* of μ_2 if and only if $\mu_2 \Rightarrow \mu_1$. Then, μ_2 is the *supermodule* of μ_1.

Lemma 1. Let μ, μ_1, μ_2 be modules so that $\mu \Leftrightarrow \mu_1 \wedge \mu_2$. Then μ_1 and μ_2 are submodules of μ.

Proof. $\mu \Leftrightarrow \mu_1 \wedge \mu_2$ infers $\mu \Rightarrow \mu_1 \wedge \mu_2$ if and only if $(\mu \Rightarrow \mu_1) \wedge (\mu \Rightarrow \mu_2)$.
$\square$

Definition 14. Let $\nu(t_1), \tilde{\nu}$ be necessities, $\rho(t_2,t_3)$ be an adaptive reaction, and $t_1 < t_2 < t_3$ be three time points.

a) $\rho(t_2,t_3)$ *fulfils* $\nu(t_1)$ if and only if $\exists \tilde{\nu}\ (\nu(t_1) \Rightarrow \tilde{\nu}) \wedge (\rho(t_2,t_3) \Rightarrow \tilde{\nu})$.

b) $\rho(t_2,t_3)$ *partially fulfils* $\nu(t_1)$ if and only if $\exists \tilde{\nu} \neq \nu(t_1)\ (\nu(t_1) \Rightarrow \tilde{\nu}) \wedge (\rho(t_2,t_3) \Rightarrow \tilde{\nu})$.

c) $\rho(t_2,t_3)$ *comletely fulfils* $\nu(t_1)$ if and only if $\rho(t_2,t_3) \Rightarrow \nu(t_1)$.

Definition 15. Let $t_1 < t_2$ be two different time points,

$\sigma_u := \exists o \in O\ \exists i \in I\ (o \cup n \setminus \{u,i\} \equiv (t_1;h)) \wedge (o \cup n \cup \{u,i\} \equiv (t_2;h))$ be a situational change, and

$\rho_i := \exists o \in O\ \exists a \in A\ (o \cup n \setminus \{i,a\} \equiv (t_1;h)) \wedge (o \cup n \cup \{i,a\} \equiv (t_2;h))$ be an adaptive reaction. Then,

a) $\tilde{\sigma}_u := \exists o \in O\ \exists i \in I\ (o \cup n \setminus \{u,i\} \equiv (t_2;h)) \wedge (o \cup n \cup \{u,i\} \equiv (t_1;h))$ is the *reverse situational change* of σ_u.

b) $\tilde{\rho}_i := \exists o \in O\ \exists a \in A\ (o \cup n \setminus \{i,a\} \equiv (t_2;h)) \wedge (o \cup n \cup \{i,a\} \equiv (t_1;h))$ is the *reverse adaptive reaction* of ρ_i.

Mathematical Multi-Level Evolution Model

Definition 16. Let $t_1<t_2$ (or $t_1>t_2$) be two time points,

$\sigma_u := \exists o \in O \, \exists i \in I \, (o \cup n \setminus \{u,i\} \, \Xi \, (t_1;h)) \wedge (o \cup n \cup \{u,i\} \, \Xi \, (t_2;h))$ be a situational change (or reverse situational change), and $\rho_i := \exists o \in O \, \exists a \in A \, (o \cup n \setminus \{i,a\} \, \Xi \, (t_1;h)) \wedge (o \cup n \cup \{i,a\} \, \Xi \, (t_2;h))$ be an adaptive reaction (or reverse adaptive reaction). Then,

a) σ_u is *partially complete* if and only if $\neg(\exists o \in O \, \exists i \in I \, (o \cup n \cup \{u,i\} \, \Xi \, (t_1;h)))$.

b) ρ_i is *partially complete* if and only if $\neg(\exists o \in O \, \exists a \in A \, (o \cup n \cup \{i,a\} \, \Xi \, (t_1;h)))$.

c) σ_u is *complete* if and only if $\neg(\exists o \in O \, \exists i \in I \, (o \cup n \setminus \{u,i\} \, \Xi \, (t_2;h)) \vee (o \cup n \cup \{u,i\} \, \Xi \, (t_1;h)))$.

d) ρ_i is *complete* if and only if $\neg(\exists o \in O \, \exists a \in A \, (o \cup n \setminus \{i,a\} \, \Xi \, (t_2;h)) \vee (o \cup n \cup \{i,a\} \, \Xi \, (t_1;h)))$.

Definition 17. Let $\sigma, \sigma_1, \sigma_2$ be situational changes, $\nu, \nu_1, \nu_2, \bar{\nu}$ be necessities, $\rho, \rho_1, \rho_2, \bar{\rho}, \bar{\rho}_1, \bar{\rho}_2$ be adaptive reactions, and $\mu := \sigma \Rightarrow \nu \Leftarrow \rho$, $\mu_1 := \sigma_1 \Rightarrow \nu_1 \Leftarrow \rho_1$, $\mu_2 := \sigma_2 \Rightarrow \nu_2 \Leftarrow \rho_2$, $\bar{\mu} := \sigma \Rightarrow \bar{\nu} \Leftarrow \bar{\rho}$ be modules, where $\bar{\nu} \neq \nu, \bar{\nu} \Rightarrow \nu, \bar{\rho} \neq \rho$.

a) μ is a *synthesis* if and only if $\exists \mu_1, \mu_2 \, \mu \Leftrightarrow \mu_1 \wedge \mu_2$.

b) μ is a *heterosynthesis* if and only if

$$\exists \mu_1, \mu_2 \left((\mu \Leftrightarrow \mu_1 \wedge \mu_2) \wedge \left(\exists \bar{\rho}_1, \bar{\rho}_2 \, ((\bar{\rho}_1 \Rightarrow \rho_1) \wedge (\bar{\rho}_2 \Rightarrow \rho_2) \wedge \exists \bar{\mu} \, (\bar{\rho}_1 \oplus \bar{\rho}_2 \Leftrightarrow \bar{\rho})) \right) \right).$$

c) μ is a *homosynthesis* if and only if

$$\exists \mu_1, \mu_2 \left((\mu \Leftrightarrow \mu_1 \wedge \mu_2) \wedge \left(\neg \exists \bar{\rho}_1, \bar{\rho}_2 \, ((\bar{\rho}_1 \Rightarrow \rho_1) \wedge (\bar{\rho}_2 \Rightarrow \rho_2) \wedge \exists \bar{\mu} \, (\bar{\rho}_1 \oplus \bar{\rho}_2 \Leftrightarrow \bar{\rho})) \right) \right).$$

Definition 18. Let $R:=\{\rho_1, \rho_2, \rho_3, ...\rho_\zeta\}$ be a set of adaptive reactions, where $\zeta \in \mathbb{N} \cup \{\infty\}$. Furthermore, let $\mu := \sigma \Rightarrow v \Leftarrow \rho$ be a module.

a) The elements of R are *compatible* with each other if and only if

$$\exists \mu \left((\rho_1 \wedge \rho_2 \wedge \rho_3 \wedge ... \wedge \rho_\zeta) \Leftrightarrow \rho\right).$$ Then, there is *compatibility* in R.

b) The elements of R are *incompatible* with each other if and only if

$$\neg \exists \mu \left((\rho_1 \wedge \rho_2 \wedge \rho_3 \wedge ... \wedge \rho_\zeta) \Leftrightarrow \rho\right).$$ Then, there is *incompatibility* in R.

APPENDIX II: DERIVATION OF THE STOCHASTIC MODEL

Module replication: We define an adaptive reaction as one evolutionary way of fulfilling the necessity for a phenotypical attribute. In addition, we term the functional sense unit associated with such an adaptive reaction, module. We use the term module as a generalization of the term gene. Thus, a module may be a gene, an allele, or a regulatory nucleotide sequence. A module can be regarded as a sequence of genomic nucleotides, which makes sense biologically. As modules are able to reproduce they can be viewed as a generalization of Dawkins' selfish genes (DAWKINS 1976).

At a time t, let the population N of $n(t)$ organisms contain $g(t)$ modules belonging to an adaptive reaction G, and $h(t)$ modules belonging to an adaptive reaction H. Let both adaptive reactions, forming $m(t)=g(t)+h(t)$ modules in N, fulfil the same necessity M. In order to simulate competition we assume $n(t)=n$ and $m(t)=m$ to be constant at any time t. Therefore, during an infinitesimal time interval dt, only the three events $M_{\uparrow\downarrow}(g(t),h(t))=$ {$g(t)$ increases by one and $h(t)$ decreaes by one}, $M_{00}(g(t),h(t))=$ {$g(t)$ and $h(t)$ remain constant}, and $M_{\downarrow\uparrow}(g(t),h(t))=$ {$g(t)$ decreases by one and $h(t)$ increases by one} are feasible; they are also mutually exclusive. Let $\mu_{\uparrow\downarrow}(g(t),h(t))$, $\mu_{00}(g(t),h(t))$, and $\mu_{\downarrow\uparrow}(g(t),h(t))$ be the corresponding probabilities:

Mathematical Multi-Level Evolution Model

$$M_{\uparrow\downarrow}(g(t), h(t)) = \{g(t + dt) = g(t) + 1 \land h(t + dt) = h(t) - 1\}$$

$$M_{00}(g(t), h(t)) = \{g(t + dt) = g(t) \land h(t + dt) = h(t)\} \quad (1)$$

$$M_{\downarrow\uparrow}(g(t), h(t)) = \{g(t + dt) = g(t) - 1 \land h(t + dt) = h(t) + 1\}$$

$$\mu_{\uparrow\downarrow}(g(t), h(t)) + \mu_{00}(g(t), h(t)) + \mu_{\downarrow\uparrow}(g(t), h(t)) = 1$$

Mathematical Multi-Level Evolution Model

In order to calculate the probabilities of formula 1 we define five additional events which mutually exclusively occur during dt: $B_G(g(t)) = \{1$ module type G produces 1 progeny module$\}$, $B_H(h(t)) = \{1$ module type H produces 1 progeny module$\}$, $D_G(g(t)) = \{1$ module type G dies$\}$, $D_H(h(t)) = \{1$ module type H dies$\}$, and $C_{GH}(g(t), h(t)) = \{0$ modules reproduce, 0 modules die$\}$. Let $b_G(g(t))$, $b_H(h(t))$, $d_G(g(t))$, $d_H(h(t))$, and $c_{GH}(g(t), h(t))$ be the corresponding probabilities. If a module of type G (or H) is born during dt, a module of either type is removed at random for m to remain constant; the probability of removing a module type H (or G) is $\frac{h(t)}{m+1}$ (or $\frac{g(t)}{m+1}$). Likewise, if a module of type G (or H) dies during dt, a module of either type is added at random for m to remain constant; the probability of adding a module type H (or G) is $\frac{h(t)}{m-1}$ (or $\frac{g(t)}{m-1}$). Therefore:

$$b_G(g(t)) + d_H(h(t)) + c_{GH}(g(t), h(t)) + d_G(g(t)) + b_H(h(t)) = 1 \quad \Rightarrow$$

$$\mu_{\uparrow\downarrow}(g(t), h(t)) = b_G(g(t)) \frac{h(t)}{m+1} + d_H(h(t)) \frac{g(t)}{m-1}$$

$$\mu^{00}(g(t), h(t)) = 1 - \mu_{\uparrow\downarrow}(g(t), h(t)) - \mu_{\downarrow\uparrow}(g(t), h(t)) \quad (2)$$

$$\mu_{\downarrow\uparrow}(g(t), h(t)) = d_G(g(t)) \frac{h(t)}{m-1} + b_H(h(t)) \frac{g(t)}{m+1}$$

Extinction probability: Let $x_L(g(0), h(0), t)$ denote the probability that module type $L \in \{G, H\}$ will have died out at time t, and if type G count is $g(0) = i$ and type H count is $h(0) = m - i = j$ at time 0. Thus, first step analysis (BRÉMAUD 1999) leads to the following

Mathematical Multi-Level Evolution Model

differential equation:

$$x_L(i,j,t+dt) = \mu_{\uparrow\downarrow}(i,j)x_L(i+1,j-1,t) + \mu_{00}(i,j)x_L(i,j,t) + \mu_{\downarrow\uparrow}(i,j)x_L(i-1,j+1,t) \Rightarrow$$

(3)

$$\dot{x}_L(i,j,t) = \frac{\mu_{\uparrow\downarrow}(i,j)}{dt}x_L(i+1,j-1,t) - \frac{\mu_{\uparrow\downarrow}(i,j)+\mu_{\downarrow\uparrow}(i,j)}{dt}x_L(i,j,t) + \frac{\mu_{\downarrow\uparrow}(i,j)}{dt}x_L(i-1,j+1,t)$$

When equilibrium occurs at an infinite time t and $\mu_{\uparrow\downarrow}(i,j) \neq 0 \wedge \mu_{\downarrow\uparrow}(i,j) \neq 0 \; \forall \, i \neq 0 \wedge j \neq 0$:

$$\lim_{t\to\infty} \dot{x}_L(i,j,t) = 0 \quad \wedge \quad x_L(i,j) = \lim_{t\to\infty} x_L(i,j,t) \quad \Rightarrow$$

$$0 = \frac{\mu_{\uparrow\downarrow}(i,j)}{dt}x_L(i+1,j-1) - \frac{\mu_{\uparrow\downarrow}(i,j)+\mu_{\downarrow\uparrow}(i,j)}{dt}x_L(i,j) + \frac{\mu_{\downarrow\uparrow}(i,j)}{dt}x_L(i-1,j+1) \Rightarrow \qquad (4)$$

$$x_L(i+1,j-1) - x_L(i,j) = \frac{\mu_{\downarrow\uparrow}(i,j)}{\mu_{\uparrow\downarrow}(i,j)}\big(x_L(i,j) - x_L(i-1,j+1)\big)$$

The partial difference equation above can be solved by summation and its repeated use:

$$\sum_{q=1}^{i-1}\big(x_L(q+1,m-q-1) - x_L(q,m-q)\big)$$

$$= \sum_{q=1}^{i-1} \frac{\mu_{\downarrow\uparrow}(q,m-q)}{\mu_{\uparrow\downarrow}(q,m-q)}\big(x_L(q,m-q) - x_L(q-1,m-q+1)\big) \qquad\Rightarrow$$

$$x_L(i,m-i) - x_L(1,m-1) \qquad\qquad (5)$$

$$= \sum_{q=1}^{i-1} \frac{\mu_{\downarrow\uparrow}(q,m-q)}{\mu_{\uparrow\downarrow}(q,m-q)} \frac{\mu_{\downarrow\uparrow}(q-1,m-q+1)}{\mu_{\uparrow\downarrow}(q-1,m-q+1)} \cdots \frac{\mu_{\downarrow\uparrow}(1,m-1)}{\mu_{\uparrow\downarrow}(1,m-1)}\big(x_L(1,m-1) - x_L(0,m)\big) \Rightarrow$$

$$x_L(1,m-1)\left(1 + \sum_{q=1}^{i-1}\prod_{r=1}^{q} \frac{\mu_{\downarrow\uparrow}(r,m-r)}{\mu_{\uparrow\downarrow}(r,m-r)}\right) = x_L(i,m-i) + x_L(0,m)\sum_{q=1}^{i-1}\prod_{r=1}^{q} \frac{\mu_{\downarrow\uparrow}(r,m-r)}{\mu_{\uparrow\downarrow}(r,m-r)}$$

Setting $i=m$ in the above equation and assuming the boundary conditions $x_G(0,m)=1$, $x_H(m,0)=1$, $x_G(m,0)=0$, and $x_H(0,m)=0$, we obtain the constants $x_G(1,m-1)$ and

Mathematical Multi-Level Evolution Model

$x_H(1, m-1)$:

$$x_G(1,m-1) = \frac{\sum_{q=1}^{m-1}\prod_{r=1}^{q}\frac{\mu_{\downarrow\uparrow}(r,m-r)}{\mu_{\uparrow\downarrow}(r,m-r)}}{1+\sum_{q=1}^{m-1}\prod_{r=1}^{q}\frac{\mu_{\downarrow\uparrow}(r,m-r)}{\mu_{\uparrow\downarrow}(r,m-r)}} \qquad (6)$$

$$x_H(1,m-1) = \frac{1}{1+\sum_{q=1}^{m-1}\prod_{r=1}^{q}\frac{\mu_{\downarrow\uparrow}(r,m-r)}{\mu_{\uparrow\downarrow}(r,m-r)}}$$

Combining the formulae 5 and 6 results in $x_G(i, m-i)$ and $x_H(i, m-i)$:

$$x_G(i,m-i) = \frac{1+\sum_{q=1}^{i-1}\prod_{r=1}^{q}\frac{\mu_{\downarrow\uparrow}(r,m-r)}{\mu_{\uparrow\downarrow}(r,m-r)}}{1+\sum_{q=1}^{m-1}\prod_{r=1}^{q}\frac{\mu_{\downarrow\uparrow}(r,m-r)}{\mu_{\uparrow\downarrow}(r,m-r)}} \sum_{q=1}^{m-1}\prod_{r=1}^{q}\frac{\mu_{\downarrow\uparrow}(r,m-r)}{\mu_{\uparrow\downarrow}(r,m-r)} - \sum_{q=1}^{i-1}\prod_{r=1}^{q}\frac{\mu_{\downarrow\uparrow}(r,m-r)}{\mu_{\uparrow\downarrow}(r,m-r)}$$

$$= \frac{\sum_{q=i}^{m-1}\prod_{r=1}^{q}\frac{\mu_{\downarrow\uparrow}(r,m-r)}{\mu_{\uparrow\downarrow}(r,m-r)}}{1+\sum_{q=1}^{m-1}\prod_{r=1}^{q}\frac{\mu_{\downarrow\uparrow}(r,m-r)}{\mu_{\uparrow\downarrow}(r,m-r)}} \qquad \wedge \qquad (7)$$

$$x_H(i,m-i) = \frac{1+\sum_{q=1}^{i-1}\prod_{r=1}^{q}\frac{\mu_{\downarrow\uparrow}(r,m-r)}{\mu_{\uparrow\downarrow}(r,m-r)}}{1+\sum_{q=1}^{m-1}\prod_{r=1}^{q}\frac{\mu_{\downarrow\uparrow}(r,m-r)}{\mu_{\uparrow\downarrow}(r,m-r)}} \qquad \Rightarrow x_G(i,m-i)+x_H(i,m-i)=1$$

Therefore, either module type G or H will certainly die out at some time.

Mean absorption time: Let $y(g(0), h(0), t)$ be the expected time either module type G or H will take to die out from the perspective of time t, if type G count is $g(0) = i$ and type H count is $h(0) = m - i = j$ at time 0. Thus, first step analysis (BRÉMAUD 1999) leads to the

Mathematical Multi-Level Evolution Model

following differential equation:

$$y(i,j,t+dt) = dt + \mu_{\uparrow\downarrow}(i,j)y(i+1,j-1,t) + \mu_{00}(i,j)y(i,j,t) + \mu_{\downarrow\uparrow}(i,j)y(i-1,j+1,t) \Rightarrow \quad (8)$$

$$\dot{y}(i,j,t) = 1 + \frac{\mu_{\uparrow\downarrow}(i,j)}{dt}y(i+1,j-1,t) - \frac{\mu_{\uparrow\downarrow}(i,j)+\mu_{\downarrow\uparrow}(i,j)}{dt}y(i,j,t) + \frac{\mu_{\downarrow\uparrow}(i,j)}{dt}y(i-1,j+1,t)$$

When equilibrium occurs at an infinite time t and $\mu_{\uparrow\downarrow}(i,j) \neq 0 \wedge \mu_{\downarrow\uparrow}(i,j) \neq 0 \; \forall \; i \neq 0 \wedge j \neq 0$:

$$\lim_{t \to \infty} \dot{y}(i,j,t) = 0 \quad \wedge \quad y(i,j) = \lim_{t \to \infty} y(i,j,t) \quad \Rightarrow$$

$$0 = 1 + \frac{\mu_{\uparrow\downarrow}(i,j)}{dt}y(i+1,j-1) - \frac{\mu_{\uparrow\downarrow}(i,j)+\mu_{\downarrow\uparrow}(i,j)}{dt}y(i,j) + \frac{\mu_{\downarrow\uparrow}(i,j)}{dt}y(i-1,j+1) \quad \Rightarrow \quad (9)$$

$$y(i,j) - y(i+1,j-1) = \frac{dt}{\mu_{\uparrow\downarrow}(i,j)} + \frac{\mu_{\downarrow\uparrow}(i,j)}{\mu_{\uparrow\downarrow}(i,j)}\big(y(i-1,j+1) - y(i,j)\big)$$

Repeated use of the above partial difference equation leads to:

$$\frac{\mu_{\uparrow\downarrow}(i,j)}{\mu_{\downarrow\uparrow}(i,j)}\big(y(i,j) - y(i+1,j-1)\big) = \frac{dt}{\mu_{\downarrow\uparrow}(i,j)} \frac{\mu_{\uparrow\downarrow}(i,j)}{\mu_{\downarrow\uparrow}(i,j)} + \big(y(i-1,j+1) - y(i,j)\big) \Rightarrow$$

$$\frac{\mu_{\uparrow\downarrow}(i,j)}{\mu_{\downarrow\uparrow}(i,j)} \frac{\mu_{\uparrow\downarrow}(i-1,j+1)}{\mu_{\downarrow\uparrow}(i-1,j+1)} \cdots \frac{\mu_{\uparrow\downarrow}(1,m-1)}{\mu_{\downarrow\uparrow}(1,m-1)}\big(y(i,j) - y(i+1,j-1)\big)$$

$$= \frac{dt}{\mu_{\uparrow\downarrow}(i,j)} \frac{\mu_{\uparrow\downarrow}(i,j)}{\mu_{\downarrow\uparrow}(i,j)} \frac{\mu_{\uparrow\downarrow}(i-1,j+1)}{\mu_{\downarrow\uparrow}(i-1,j+1)} \cdots \frac{\mu_{\uparrow\downarrow}(1,m-1)}{\mu_{\downarrow\uparrow}(1,m-1)} \quad (10)$$

$$+ \frac{dt}{\mu_{\uparrow\downarrow}(i-1,j+1)} \frac{\mu_{\uparrow\downarrow}(i-1,j+1)}{\mu_{\downarrow\uparrow}(i-1,j+1)} \frac{\mu_{\uparrow\downarrow}(i-2,j+2)}{\mu_{\downarrow\uparrow}(i-2,j+2)} \cdots \frac{\mu_{\uparrow\downarrow}(1,m-1)}{\mu_{\downarrow\uparrow}(1,m-1)}$$

$$+ \cdots + \frac{dt}{\mu_{\uparrow\downarrow}(1,m-1)} \frac{\mu_{\uparrow\downarrow}(1,m-1)}{\mu_{\downarrow\uparrow}(1,m-1)} + \big(y(0,m) - y(1,m-1)\big)$$

Mathematical Multi-Level Evolution Model

Assuming the boundary condition $y(0,m)=0$ and subsequent summation results in:

$$\left(y(i,m-i)-y(i+1,m-i-1)\right)\prod_{r=1}^{i}\frac{\mu_{\uparrow\downarrow}(r,m-r)}{\mu_{\downarrow\uparrow}(r,m-r)}$$

$$=\sum_{r=1}^{i}\frac{dt}{\mu_{\uparrow\downarrow}(r,m-r)}\prod_{s=1}^{r}\frac{\mu_{\uparrow\downarrow}(s,m-s)}{\mu_{\downarrow\uparrow}(s,m-s)}-y(1,m-1) \quad\Rightarrow$$

$$\sum_{q=1}^{i-1}\left(y(q,m-q)-y(q+1,m-q-1)\right)$$

$$=\sum_{q=1}^{i-1}\frac{\sum_{r=1}^{q}\dfrac{dt}{\mu_{\uparrow\downarrow}(r,m-r)}\prod_{s=1}^{r}\dfrac{\mu_{\uparrow\downarrow}(s,m-s)}{\mu_{\downarrow\uparrow}(s,m-s)}-y(1,m-1)}{\prod_{r=1}^{q}\dfrac{\mu_{\uparrow\downarrow}(r,m-r)}{\mu_{\downarrow\uparrow}(r,m-r)}} \quad\Rightarrow$$

$$y(1,m-1)-y(i,m-i)+y(1,m-1)\sum_{q=1}^{i-1}\prod_{r=1}^{q}\frac{\mu_{\downarrow\uparrow}(r,m-r)}{\mu_{\uparrow\downarrow}(r,m-r)} \tag{11}$$

$$=\sum_{q=1}^{i-1}\prod_{r=1}^{q}\frac{\mu_{\downarrow\uparrow}(r,m-r)}{\mu_{\uparrow\downarrow}(r,m-r)}\sum_{r=1}^{q}\frac{dt}{\mu_{\uparrow\downarrow}(r,m-r)}\prod_{s=1}^{r}\frac{\mu_{\uparrow\downarrow}(s,m-s)}{\mu_{\downarrow\uparrow}(s,m-s)}$$

Setting $i=m$ and assuming the boundary condition $y(m,0)=0$ we acquire the constant $y(1,m-1)$:

$$y(1,m-1)=\frac{\sum_{q=1}^{m-1}\prod_{r=1}^{q}\dfrac{\mu_{\downarrow\uparrow}(r,m-r)}{\mu_{\uparrow\downarrow}(r,m-r)}\sum_{r=1}^{q}\dfrac{dt}{\mu_{\uparrow\downarrow}(r,m-r)}\prod_{s=1}^{r}\dfrac{\mu_{\uparrow\downarrow}(s,m-s)}{\mu_{\downarrow\uparrow}(s,m-s)}}{1+\sum_{q=1}^{m-1}\prod_{r=1}^{q}\dfrac{\mu_{\downarrow\uparrow}(r,m-r)}{\mu_{\uparrow\downarrow}(r,m-r)}} \tag{12}$$

Combining the formulae 11 and 12 we obtain $y(i, m-i)$:

$$y(i,m-i) = \frac{1+\sum_{q=1}^{i-1}\prod_{r=1}^{q}\frac{\mu_{\downarrow\uparrow}(r,m-r)}{\mu_{\uparrow\downarrow}(r,m-r)}}{1+\sum_{q=1}^{m-1}\prod_{r=1}^{q}\frac{\mu_{\downarrow\uparrow}(r,m-r)}{\mu_{\uparrow\downarrow}(r,m-r)}}\sum_{q=1}^{m-1}\prod_{r=1}^{q}\frac{\mu_{\downarrow\uparrow}(r,m-r)}{\mu_{\uparrow\downarrow}(r,m-r)}\sum_{r=1}^{q}\frac{dt}{\mu_{\uparrow\downarrow}(r,m-r)}\prod_{s=1}^{r}\frac{\mu_{\uparrow\downarrow}(s,m-s)}{\mu_{\downarrow\uparrow}(s,m-s)}$$

$$-\sum_{q=1}^{i-1}\prod_{r=1}^{q}\frac{\mu_{\downarrow\uparrow}(r,m-r)}{\mu_{\uparrow\downarrow}(r,m-r)}\sum_{r=1}^{q}\frac{dt}{\mu_{\uparrow\downarrow}(r,m-r)}\prod_{s=1}^{r}\frac{\mu_{\uparrow\downarrow}(s,m-s)}{\mu_{\downarrow\uparrow}(s,m-s)}$$

(13)

Asexual reproduction: Let each of n asexual organisms in a population N contain either one module type G or one module type H. Consequently, the total number of modules is $m=n$, and type G organism count is $g(t)$ and type H organism count is $h(t)$. Furthermore, the probability that in N one organism of type G (or H) is born at time t during dt equals $b_G(g(t))$ (or $b_H(h(t))$). Likewise, the probability that in N one organism of type G (or H) dies at time t during dt equals $d_G(g(t))$ (or $d_H(h(t))$). Let $b_G = b_G(1)$ (or $b_H = b_H(1)$) be the probability of one organism type G (or H) to produce one progeny, and let $d_G = d_G(1)$ (or $d_H = d_H(1)$) be the probability of one organism type G (or H) to die during any infinitesimal time interval dt:

$$b_G(g(t)) = g(t)b_G(1) = g(t)b_G$$

$$b_H(h(t)) = h(t)b_H(1) = h(t)b_H$$

(14)

$$d_G(g(t)) = g(t)d_G(1) = g(t)d_G$$

$$d_H(h(t)) = h(t)d_H(1) = h(t)d_H$$

Sexual reproduction: Let each of n sexual organisms in a population N contain a combination of two modules G or H, where each module within a combination is on a different chromosome. Consequently, the total number of modules is $m=2n$, and the population N consists of three organism types: the homozygote GG, the heterozygote GH, and the homozygote HH. Let b_{GG}, b_{GH}, and b_{HH} be the corresponding probabilities of one organism to produce one progeny during any infinitesimal time interval dt. Moreover, let d_{GG}, d_{GH}, and d_{HH} be the corresponding probabilities of one organism to die during dt. These probabilities, which refer to organisms, determine the probabilities of individual modules to die or to duplicate during dt: $b_G(1)$, $b_H(1)$, $d_G(1)$, and $d_H(1)$. These probabilities, which refer to modules, depend on the type of the corresponding partner modules. Let $p_L(G, g(t))$ (or $p_L(H, h(t))$) be the probability that a module of type $L \in \{G, H\}$ has a partner module of type G (or H), given that the partner module count is $g(t)$ (or $h(t)$):

$$p_G(G, g(t)) = \frac{g(t)-1}{m-1}$$

$$p_G(H, h(t)) = \frac{h(t)}{m-1}$$

$$p_H(G, g(t)) = \frac{g(t)}{m-1}$$

$$p_H(H, h(t)) = \frac{h(t)-1}{m-1}$$

(15)

In order to simplify the sexual reproduction process we assume that one organism, *i.e.* two modules, can only produce one gamete, *i.e.* one module, at time t during dt. Thus, the probability for a module to duplicate within a reproducing organism is $\frac{1}{2}$:

$$b_G(g(t)) = g(t)b_G(1) = g(t)\left(\frac{p_G(G,g(t))}{2}b_{GG} + \frac{p_G(H,m-g(t))}{2}b_{GH}\right)$$

$$= \frac{g(t)}{2(m-1)}\left(g(t)b_{GG} - b_{GG} + mb_{GH} - g(t)b_{GH}\right)$$

$$b_H(h(t)) = h(t)b_H(1) = h(t)\left(\frac{p_H(H,h(t))}{2}b_{HH} + \frac{p_H(G,m-h(t))}{2}b_{GH}\right) \quad (16)$$

$$= \frac{h(t)}{2(m-1)}\left(h(t)b_{HH} - b_{HH} + mb_{GH} - h(t)b_{GH}\right)$$

When an organism dies both of its modules G or H die simultaneously. Therefore, the probability for a module to disappear within a dying organism is 1:

$$d_G(g(t)) = g(t)d_G(1) = g(t)\left(p_G(G,g(t))d_{GG} + p_G(H,m-g(t))d_{GH}\right)$$

$$= \frac{g(t)}{m-1}\left(g(t)d_{GG} - d_{GG} + md_{GH} - g(t)d_{GH}\right)$$

$$\quad (17)$$

$$d_H(h(t)) = h(t)d_H(1) = h(t)\left(p_H(H,h(t))d_{HH} + p_H(G,m-h(t))d_{GH}\right)$$

$$= \frac{h(t)}{m-1}\left(h(t)d_{HH} - d_{HH} + md_{GH} - h(t)d_{GH}\right)$$

Fitness and fluctuation: We define the fitness $f_W(n)$ of a population containing $n \in \{1,2,3,...\}$ organisms of type $W \in \{G, H, GG, GH, HH\}$ as the expectation of its increase in organism count per fixed generation time Δt. Furthermore, we define the fluctuation $v_W(n)$ of a population containing n organisms of type W as the variance of its increase in organism count per fixed generation time Δt. Specifically, fitness $f_W = f_W(1)$ or fluctuation $v_W = v_W(1)$ of an individual organism equals fitness or fluctuation of a population containing only this organism. We set $\Delta t = dt$ to be infinitesimal short. The events $B_W(1)$ and $D_W(1)$ are stochastically independent. Thus:

$$f_W = \frac{b_W - d_W}{dt} \quad \wedge \quad v_w(1) = \frac{b_W + d_W}{dt} \quad \Rightarrow$$

$$b_W = \frac{v_W + f_W}{2} dt \quad \wedge \quad d_W(1) = \frac{v_W - f_W}{2} dt \tag{18}$$

$$b_W, d_W \geq 0 \quad \Rightarrow \quad -v_W \leq f_W \leq v_W$$

SUPPORTING CHARTS AND GRAPHS

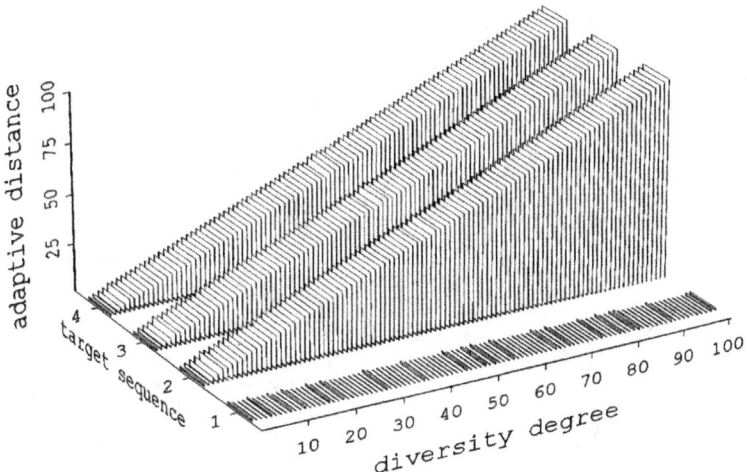

FIGURE 1. – 100 allopatric speciation experiments after 50,000 generations. With random start populations, each experiment was performed in a different ecological niche, where a set of four target sequences had a different diversity degree. The adaptive reactions and subreactions of these target sequences were isoadditive. At each diversity degree the adaptive distances from the population to the target sequences are arranged in ascending order so that target sequence 1 shows the lowest adaptive distance, target sequence 4 the highest. The same results were obtained in 100 allopatric speciation experiments with selected start populations.

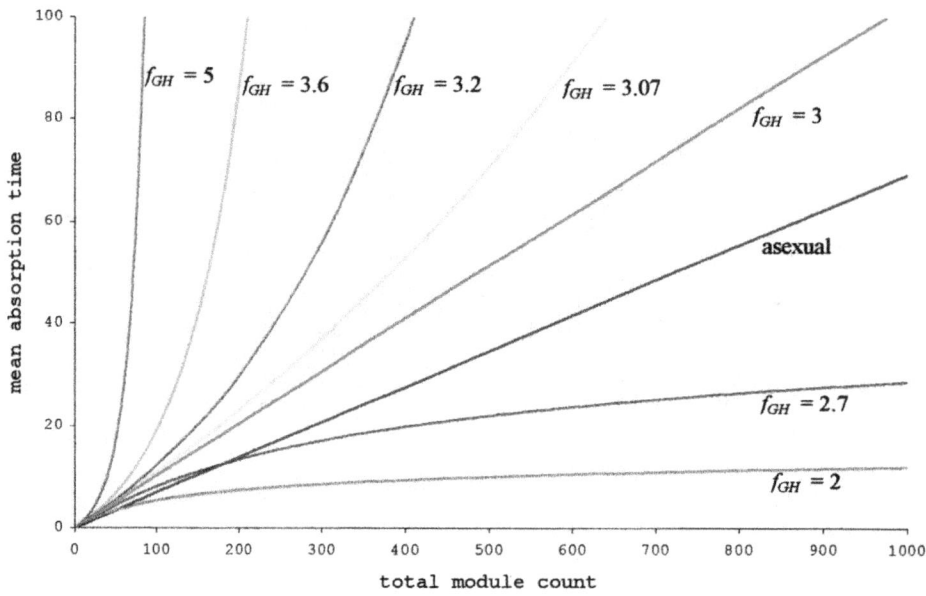

FIGURE 2. – **Mean absorption time $y(i,j)$ when G is as fit as H and $i=j$**. This relates to the total module count $m = 2, 4, 6, ..., 1000$ in sexual and asexual species. In each species a population consists of two types of modules, G and H. This leads to three kinds of organisms GG, GH, and HH in sexual species or two kinds of organisms G and H in asexual species. Each organism has the same fluctuation $v_G = v_{GG} = v_{GH} = v_{HH} = v_H = 10$. Each homozygote and asexual organism has the same fitness $f_G = f_{GG} = f_{HH} = f_H = 3$. In different populations, the fitness values f_{GH} of heterozygotes range from 2 to 5. When $f_{GH} = 3$ both module types are isoadditive, when $f_{GH} < 3$ they are hypoadditive, and when $f_{GH} > 3$ they are hyperadditive. At the start module type G count i and module type H count j are equal, *i.e.* $i = j$.

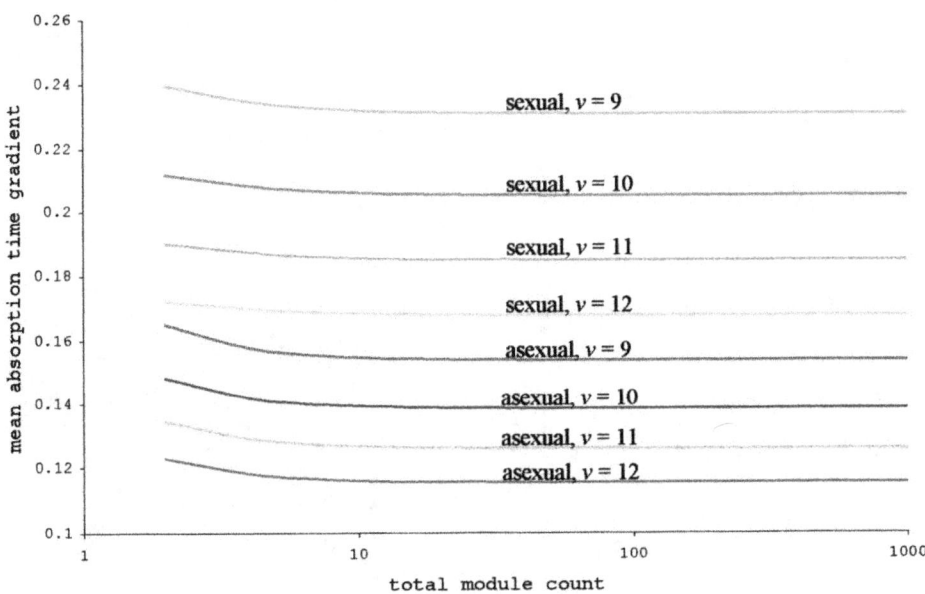

FIGURE 3. – **Mean absorption time gradient** $\Delta_{i,j} y(i,j) = y(i+1, j+1) - y(i,j)$ **when G is as fit as H and $i=j$.** This relates to the total module count $m = 2, 4, 6, ..., 1000$ in sexual and asexual species. In each species a population consists of two types of modules, G and H. This leads to three kinds of organisms GG, GH, and HH in sexual species or two kinds of organisms G and H in asexual species. Each organism has the same fitness $f_G = f_{GG} = f_{GH} = f_{HH} = f_H = 3$. In different populations, the fluctuations $v = v_G = v_{GG} = v_{GH} = v_{HH} = v_H$ range from 9 to 12. At the start module type G count i and module type H count j are equal, i.e. $i = j$.

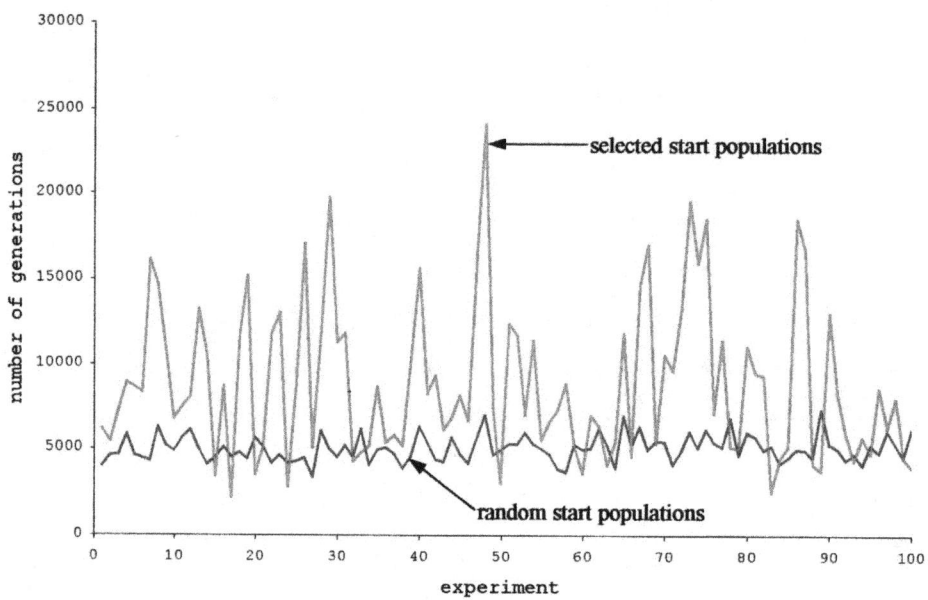

FIGURE 4. – Required number of generations t for allopatric speciation. In 200 computer experiments, a set of four target sequences with isoadditive adaptive reactions and subreactions had diversity degree 75. 100 experiments were with random start populations and 100 with selected start populations. With the random start populations, the average t was $A_t = 5059.5$, and with the selected start populations, $A_t = 8882.1$.

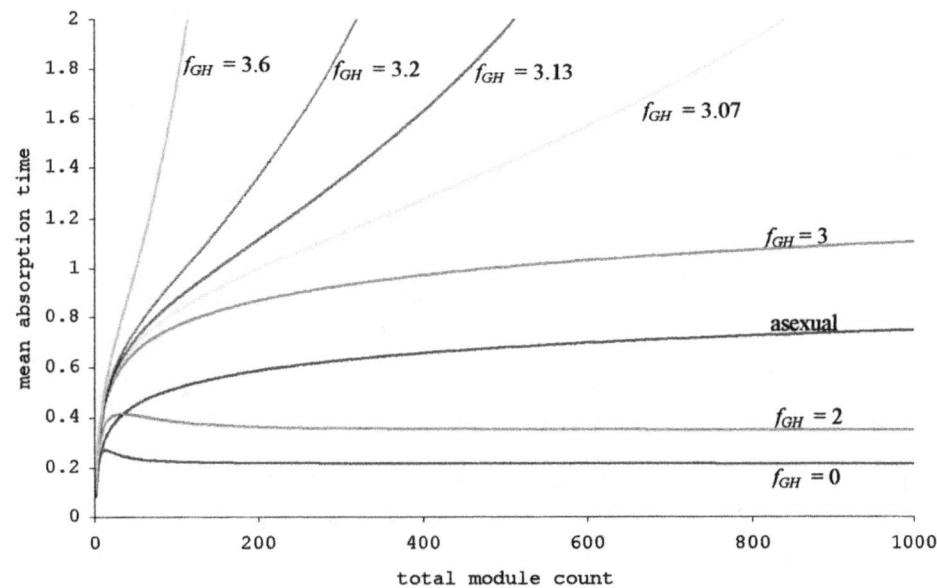

FIGURE 5. – **Mean absorption time $y(i,j)$ when G is as fit as H and $i=1$.** This relates to the total module count $m = 2, 4, 6, \ldots, 1000$ in sexual and asexual species. In each species a population consists of two types of modules, G and H. This leads to three kinds of organisms GG, GH, and HH in sexual species or two kinds of organisms G and H in asexual species. Each organism has the same fluctuation $v_G = v_{GG} = v_{GH} = v_{HH} = v_H = 10$. Each homozygote and asexual organism has the same fitness $f_G = f_{GG} = f_{HH} = f_H = 3$. In different populations, the fitness values f_{GH} of heterozygotes range from 0 to 3.6. When $f_{GH} = 3$ both module types are isoadditive, when $f_{GH} < 3$ they are hypoadditive, and when $f_{GH} > 3$ they are hyperadditive. At the start module type G count is $i = 1$ and module type H count is $j = m - 1$.

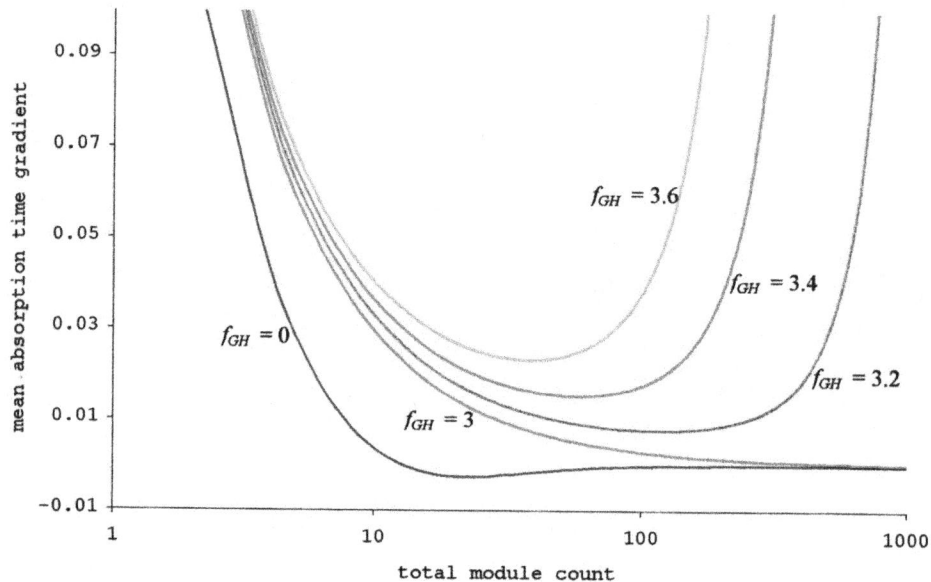

FIGURE 6. – Mean absorption time gradient $\Delta_j y(1,j) = y(1,j+2) - y(1,j)$ when G is as fit as H and $i=1$. This relates to the total module count $m = 2, 4, 6, \ldots, 1000$ in sexual species. In each species a population consits of two types of modules, G and H, leading to three kinds of organisms GG, GH, and HH. Each organism has the same fluctuation $v_{GG} = v_{GH} = v_{HH} = 10$. Each homozygote has the same fitness $f_{GG} = f_{HH} = 3$. In different populations, the fitness values f_{GH} of heterozygotes range from 0 to 3.6. When $f_{GH} = 3$ both module types are isoadditive, when $f_{GH} < 3$ they are hypoadditive, and when $f_{GH} > 3$ they are hyperadditive. At the start module type G count is $i = 1$ and module type H count is $j = m - 1$.

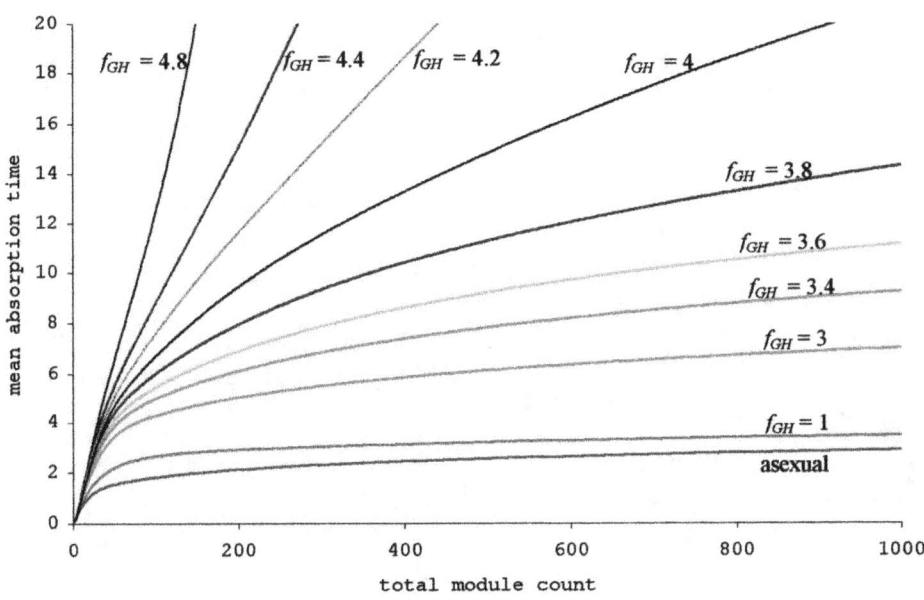

FIGURE 7. – **Mean absorption time $y(i,j)$ when G is fitter than H and $i=j$.** This relates to the total module count $m = 2, 4, 6, \ldots, 1000$ in sexual and asexual species. In each species a population consists of two types of modules, G and H. This leads to three kinds of organisms GG, GH, and HH in sexual species or two kinds of organisms G and H in asexual species. Each organism has the same fluctuation $v_G = v_{GG} = v_{GH} = v_{HH} = v_H = 10$. The homozygotes and asexual organisms with G have fitness $f_{GG} = f_G = 4$; the homozygotes and asexual organisms with H have fitness $f_{HH} = f_H = 2$. In different populations, the fitness values f_{GH} of heterozygotes range from 1 to 4.8. When $f_{GH} = 4$ both module types are isoadditive, when $f_{GH} < 4$ they are hypoadditive, and when $f_{GH} > 4$ they are hyperadditive. At the start module type G count i and module type H count j are equal, *i.e.* $i = j$.

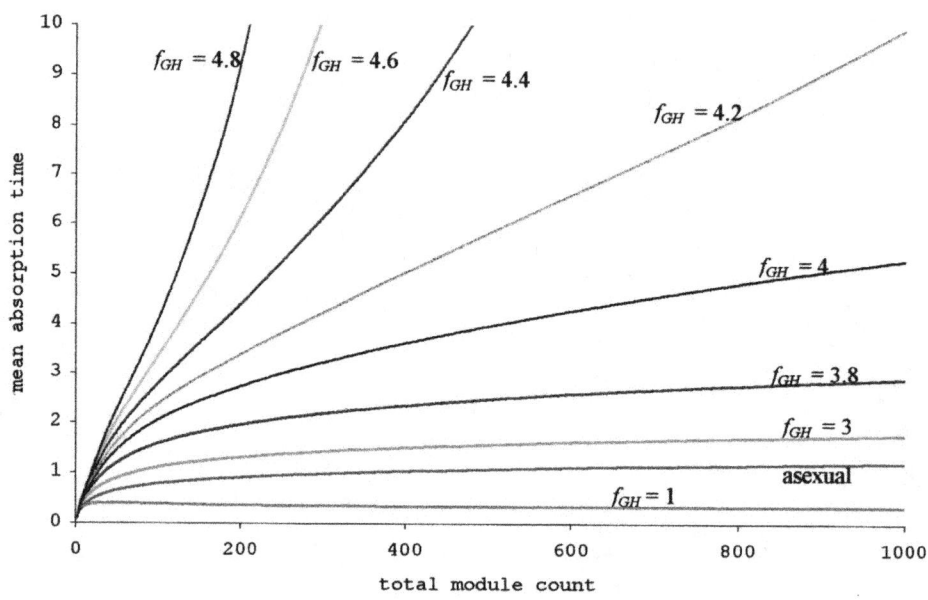

FIGURE 8. – **Mean absorption time $y(i,j)$ when G is fitter than H and $i=1$.** This relates to the total module count $m = 2, 4, 6, \ldots, 1000$ in sexual and asexual species. In each species a population consists of two types of modules, G and H. This leads to three kinds of organisms GG, GH, and HH in sexual species or two kinds of organisms G and H in asexual species. Each organism has the same fluctuation $v_G = v_{GG} = v_{GH} = v_{HH} = v_H = 10$. The homozygotes and asexual organisms with G have fitness $f_{GG} = f_G = 4$; the homozygotes and asexual organisms with H have fitness $f_{HH} = f_H = 2$. In different populations, the fitness values f_{GH} of heterozygotes range from 1 to 4.8. When $f_{GH} = 4$ both module types isoadditive, when $f_{GH} < 4$ they are hypoadditive, and when $f_{GH} > 4$ they are hyperadditive. At the start module type G count is $i = 1$ and module type H count is $j = m - 1$.

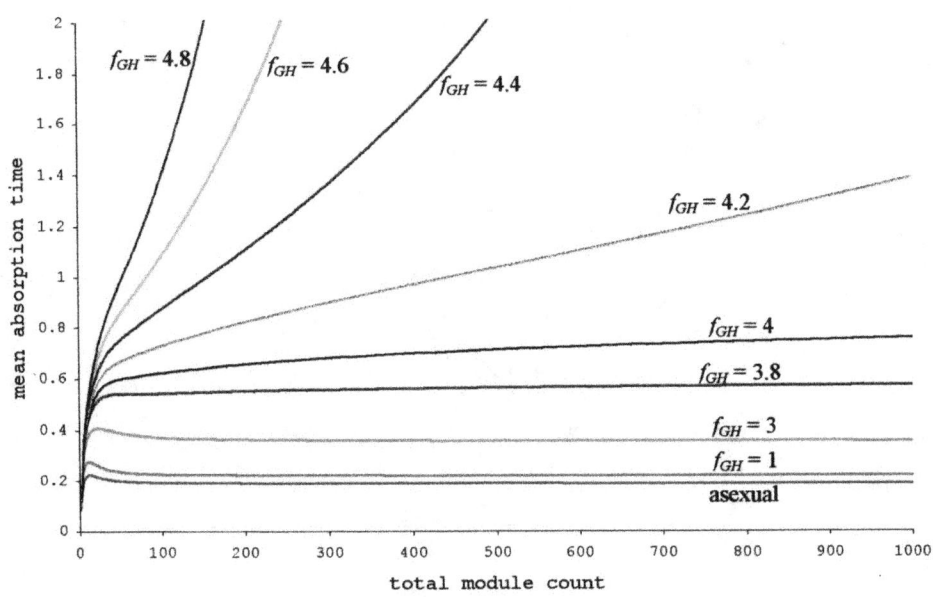

FIGURE 9. – Mean absorption time $y(i,j)$ when G is fitter than H and $j = 1$. This relates to the total module count $m = 2, 4, 6, \ldots, 1000$ in sexual and asexual species. In each species a population consists of two types of modules, G and H. This leads to three kinds of organisms GG, GH, and HH in sexual species or two kinds of organisms G and H in asexual species. Each organism has the same fluctuation $v_G = v_{GG} = v_{GH} = v_{HH} = v_H = 10$. The homozygotes and asexual organisms with G have fitness $f_{GG} = f_G = 4$; the homozygotes and asexual organisms with H have fitness $f_{HH} = f_H = 2$. In different populations, the fitness values f_{GH} of heterozygotes range from 1 to 4.8. When $f_{GH} = 4$ both module types are isoadditive, when $f_{GH} < 4$ they are hypoadditve, and when $f_{GH} > 4$ they are hyperadditve. At the start module type G count is $i = m - 1$ and module type H count is $j = 1$.

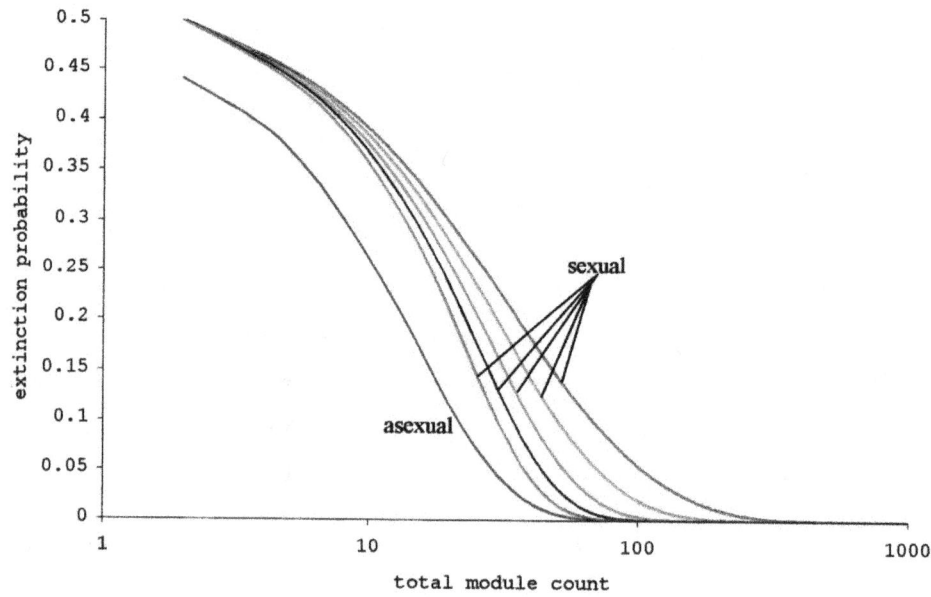

FIGURE 10. – **Extinction probability $x_G(i,j)$ of the fittest module type G when $i=j$.** This relates to the total module count $m = 2, 4, 6, \ldots, 1000$ in sexual and asexual species. In each species a population consists of two types of modules, G and H. This leads to three kinds of organisms GG, GH, and HH in sexual species or two kinds of organisms G and H in asexual species. Each organism has the same fluctuation $v_G = v_{GG} = v_{GH} = v_{HH} = v_H = 10$. The homozygotes and asexual organisms with G have fitness $f_{GG} = f_G = 4$; the homozygotes and asexual organisms with H have fitness $f_{HH} = f_H = 2$. In different populations, the fitness values f_{GH} of heterozygotes are 1, 2, 3, 4, 5 (sequence of curves runs from right to left). At the start module type G count i and module type H count j are equal, i.e. $i=j$. When $x_G(i,j) > 0$, an uncertainty principle can be seen in the corresponding module counts m.

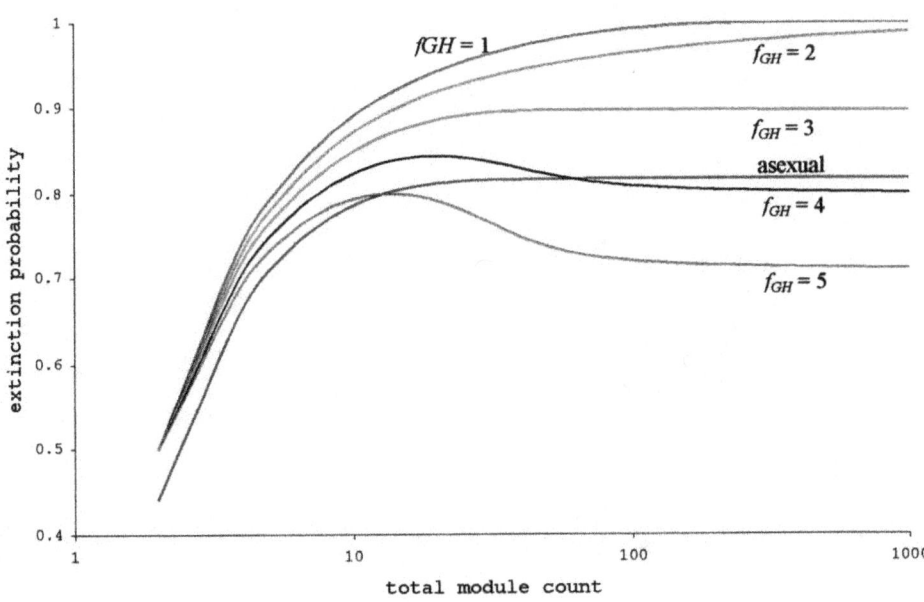

FIGURE 11. – Extinction probability $x_G(i,j)$ of the fittest module type G when $i=1$. This relates to the total module count $m = 2, 4, 6, ..., 1000$ in sexual and asexual species. In each species a population consists of two types of modules, G and H. This leads to three kinds of organisms GG, GH, and HH in sexual species or two kinds of organisms G and H in asexual species. Each organism has the same fluctuation $v_G = v_{GG} = v_{GH} = v_{HH} = v_H = 10$. The homozygotes and asexual organisms with G have fitness $f_{GG} = f_G = 4$; the homozygotes and asexual organisms with H have fitness $f_{HH} = f_H = 2$. In different populations, the fitness values f_{GH} of heterozygotes range from 1 to 5. At the start module type G count is $i=1$ and module type H count is $j = m - 1$. An uncertainty principle, i.e. $x_G(i,j) > 0$, can be seen in all module counts m.

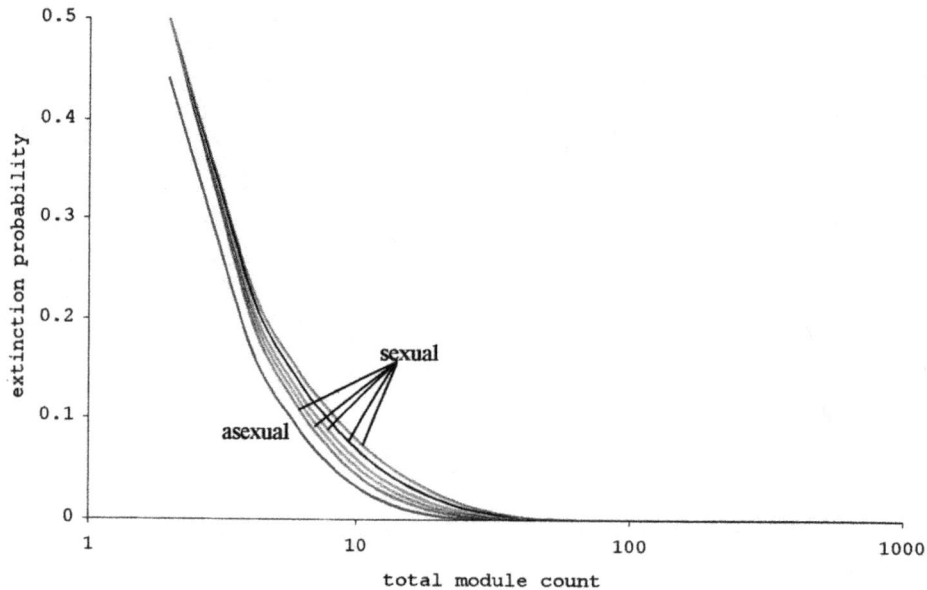

FIGURE 12. – Extinction probability $x_G(i,j)$ of the fittest module type G when $j=1$. This relates to the total module count $m = 2, 4, 6, \ldots, 1000$ in sexual and asexual species. In each species a population consists of two types of modules, G and H. This leads to three kinds of organisms GG, GH, and HH in sexual species or two kinds of organisms G and H in asexual species. Each organism has the same fluctuation $v_G = v_{GG} = v_{GH} = v_{HH} = v_H = 10$. The homozygotes and asexual organisms with G have fitness $f_{GG} = f_G = 4$; the homozygotes and asexual organisms with H have fitness $f_{HH} = f_H = 2$. In different populations, the fitness values f_{GH} of heterozygotes are 5, 4, 3, 2, 1 (sequence of curves runs from right to left). At the start module type G count is $i = m - 1$ and module type H count is $j = 1$. When $x_G(i,j) > 0$, an uncertainty principle can be seen in the corresponding module counts m.

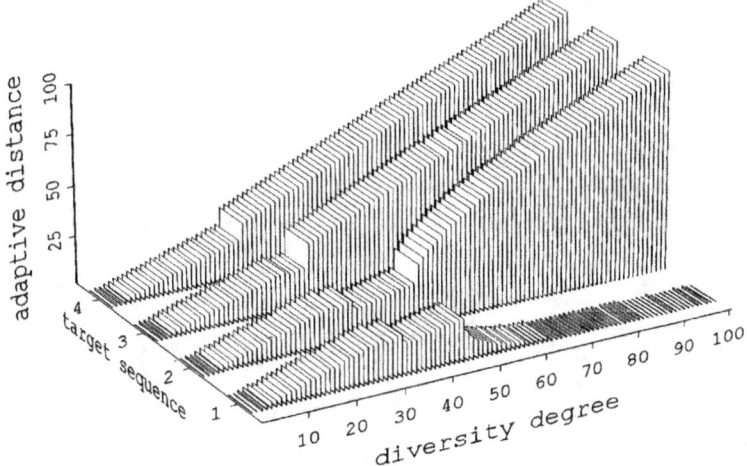

FIGURE 13. – 100 sympatric speciation experiments after 50,000 generations. With random start populations, each experiment was performed in a different ecological niche, where a set of four target sequences had a different diversity degree. The adaptive reactions and subreactions of these target sequences were hyperadditive. At each diversity degree the adaptive distances from the population to the target sequences are arranged in ascending order so that target sequence 1 shows the lowest adaptive distance, target sequence 4 the highest. Similar results were obtained in 100 sympatric speciation experiments with selected start populations.

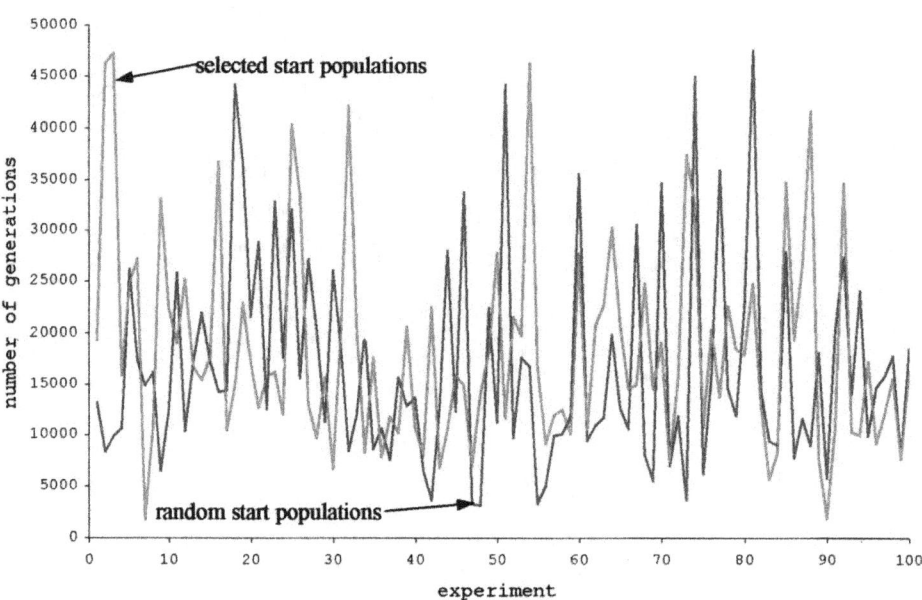

FIGURE 14. – **Required number of generations t for sympatric speciation.** In 200 computer experiments, a set of four targert sequences with hyperadditive adaptive reactions and subreactions had diversity degree 75. 100 experiments were with random start populations and 100 with selected start populations. With the random start populations, the average t was $A_t = 16,763.1$, and with the selected start populations, $A_t = 18,456.2$.

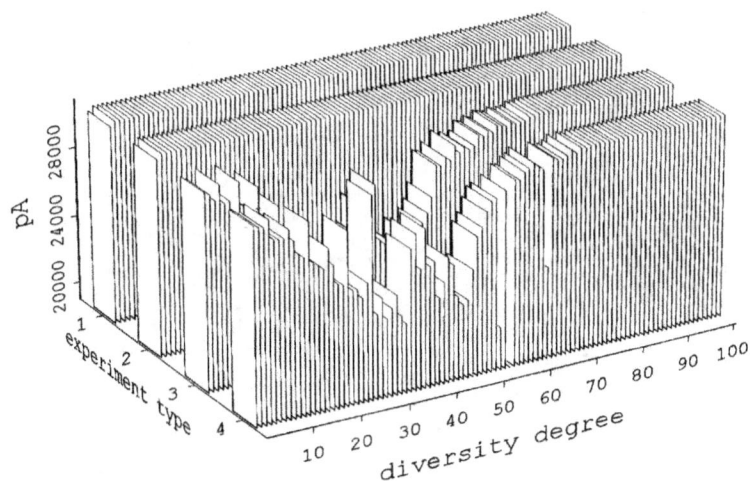

FIGURE 15. – Final pA values in the computer experiments. Experiment type 1 with isoadditive adaptive reactions and random start populations, type 2 with isoadditive adaptive reactions and selected start populations, type 3 with hyperadditive adaptive reactions and random start populations, type 4 with hyperadditive adaptive reactions and selected start populations.

www.ingramcontent.com/pod-product-compliance
Lightning Source LLC
Chambersburg PA
CBHW082216220526

45470CB00010B/3188